SpringerBriefs in Environmental Science

T0216077

For further volumes:
http://www.springer.com/series/8868

Charles Clarence Morris
Thomas Paul Simon

Nutrient Indicator Models for Determining Biologically Relevant Levels

A Case Study Based on the Corn Belt and Northern Great Plain Nutrient Ecoregion

 Springer

Charles Clarence Morris
Department of Biology
Indiana State University
Chestnut Street 600
Terre Haute
IN 47809
USA

Prof. Thomas Paul Simon
Department of Biology
Indiana State University
Chestnut Street 600
Terre Haute
IN 47809
USA

ISSN 2191-5547
ISBN 978-94-007-4128-7
DOI 10.1007/978-94-007-4129-4
Springer Dordrecht Heidelberg New York London

e-ISSN 2191-5555
e-ISBN 978-94-007-4129-4

Library of Congress Control Number: 2012931950

Printed on acid-free paper

Springer is part of Springer Science+Business Media (www.springer.com)

Contents

Abbreviations

A	Amps
ANOVA	Analysis of variance
°C	Degrees celsius
CCC	Criterion continuous concentration
EPA	Environmental Protection Agency
ft/s	Feet per second
HSD	Honest significant difference
IBI	Index of biotic integrity
l	Liter
m	Meters
mm	Millimeters
mg	Milligrams
mi	Mile
mg/l	Milligrams per liter
ml	Milliliter
MS	Matrix spike
MSDs	Matrix spike duplicate
NBI	Nutrient biotic index
NTU	Nephelometric turbidity unit
ppm	Parts per million
s	Seconds
sq	Square
SRI	Shift response interval
TKN	Total kjeldahl nitrogen
TN	Total nitrogen
TP	Total phosphorus
TRI	Test response interval
μs	Microsiemens
V	Volts
W	Watts
3D	Three dimensional

Nutrient Indicator Models for Determining Biologically Relevant Levels: A Case Study Based on the Corn Belt and Northern Great Plain Nutrient Ecoregion

Abstract The complex interactions between forms of Nitrogen and Phosphorus require development of regional nutrient threshold models. Our objectives included the development of a biotic model capable of distinguishing contributions of various nutrients in streams fish assemblages. A second objective was to establish an approach for designating defensible nutrient biotic index (NBI) score thresholds and corresponding nutrient concentrations, above which fish assemblages show alterations. Nutrient and fish assemblage data collected from 1274 reaches between 1996 and 2007 from the Corn Belt and Northern Great Plain Nutrient Ecoregion were reviewed for outliers, sorted into three drainage class groups, and arranged into 15 ranges or "bins" using the Jenks optimization method to calculate nutrient specific fish species tolerance scores. These scores where then used to generate Nutrient Biotic Indexes (NBI) for identifying fish assemblage response mechanisms. We observed a single break point for unionized ammonia, with an $NBI_{Unionized\ Ammonia}$ score shift occurring at a mean concentration of 0.03 mg/L. Three break points were observed for Nitrogen, Nitrate + Nitrite, demonstrating significant $NBI_{Nitrate+Nitrite}$ score shifts at concentrations of 1.09, 3.15 and 6.87 mg/L respectively. The observed relationship produced a convex curve suggesting an enrichment signature. Two break points were observed for Total Kjeldahl Nitrogen (TKN) at mean concentrations of 0.68 and 1.27 mg/L respectively. One significant break point was observed for TN at a mean concentration of 3.30 mg/L. One significant break point was observed for TP at a mean concentration of TP 0.32 mg/L. One significant break point was observed for Chlorophyll a (periphyton) at a mean concentration of TP 134.14 mg/m². Two significant break points were observed for Chlorophyll a (phytoplankton), which occurred at concentrations of 10.98 and 49.13 µg/L, respectively. Proposed mean protection values are 3.0 µg/L for Unionized Ammonia, 130 µg/L for Nitrogen, Nitrate + Nitrite, 40 µg/L for TKN, 70 µg/L for TP, and 2.33 µg/L for Chlorophyll a (phytoplankton). Criteria established at or below these benchmarks should protect for both biological integrity of fish assemblages, as well as limit nutrient loadings causing dead zones.

C. C. Morris and T. P. Simon, *Nutrient Indicator Models for Determining Biologically Relevant Levels*, SpringerBriefs in Environmental Science, DOI: 10.1007/978-94-007-4129-4_1, © The Author(s) 2012

Keywords Nutrient models · Species optima · Nutrient biotic indices · Test response intervals · Shift response intervals

1 Introduction

There is a growing need to understand the sources and transport of both Nitrogen and Phosphorus and their complex interactions in efforts to develop effective nutrient management (Boesch 2002; Rabalias et al. 2002; Howarth and Marino 2006; Scavia and Donnelly 2007). The transport of nutrients into coastal marine waters is of particular concern due to eutrophication and hypoxia in coastal estuaries (Howarth et al. 2002; Mississippi River/Gulf of Mexico Nutrient Task Force 2004). Alexander et al. (2008) estimates that the Corn Belt (Midwestern United States) is responsible for about 60% of the Nitrogen and 54% of the Phosphorus load delivered to the northern Gulf of Mexico. Recent evidence suggests that persistent hypoxia in the Gulf of Mexico has caused an ecological shift, such that the ecosystem has become more sensitive to nutrient loads (Turner et al. 2008). In addition, enrichment of surface waters by Nitrogen and Phosphorus has contributed to the impairment of about half of the freshwater resources of the United States (Gibson et al. 2000a, b; Reckhow et al. 2005). Thus, in order to affect Gulf hypoxia and nutrient loading in freshwater systems, management changes are needed in the Corn Belt.

The amount of nutrients entering a stream is determined both by the supply size, which is primarily influenced by land-use characteristics, and the delivery efficiency of hydrologic pathways connecting uplands to drainage networks. The amount of nutrients delivered to downstream sinks is controlled by both the permanent removal in sediments and temporary storage capacity of the system. Nitrogen and Phosphorus enter streams from point and nonpoint sources (e.g., drainage tile, runoff, groundwater, atmospheric deposition) (Omernik 1976). Large nutrient sources are typically associated with agricultural activities; however, in some regions atmospheric deposition from fossil fuel combustion can be substantial (Boyer et al. 2002), as well as inputs from septic systems, leaking sewers, and wastewater treatment plants (Brakebill and Preston 2004; Wollheim et al. 2005). Dodds and Welch (2000) indicated that due to the complex interactions between the various nutrient forms, Nitrogen and Phosphorus can no longer be accepted as sole limiting factors in either marine or freshwaters. Nitrate is the predominant form of Nitrogen in many streams, because it is highly soluble and readily leached from soils. Likewise, Ammonium is also common, but less prevalent in the water column because it is readily immobilized, adsorbs to negatively charged clay particles and organic matter, and is often nitrified in small streams. Total Kjeldahl Nitrogen (TKN) is the sum of organic Nitrogen, Ammonia and Ammonium. Dissolved or particulate organic Nitrogen may also be present in substantial amounts in some streams (Kaushal and Lewis 2005).

The U.S. Environmental Protection Agency (EPA) (1998a) has recommended a national strategy for establishing nutrient criteria. Nine states, including Indiana, are identified as contributing 75% of the Nitrogen and Phosphorus delivery to the Gulf of Mexico (Alexander et al. 2008). Natural soil enrichment and different precipitation patterns require that the typical single criterion consensus approach for nutrient criteria be established regionally to take into consideration geographical and climatological variation (Omernik 1977). For each nutrient region, recommendations for TN and TP, as causal variables, and Chlorophyll a and turbidity, as early indicator response variables, are recommended (Gibson et al. 2000a, b). Secondary response variables include aquatic biological assemblages.

Since aquatic assemblages are widely distributed and have a favorable public perception, they have been used extensively as indicators of organic enrichment due to response patterns in species composition, and differential sensitivity to environmental impact. Eutrophication indicators, such as the saprobien index (Chutter 1972), Hilsenhoff biotic index (Hilsenhoff 1987), and the nutrient biotic index (Smith et al. 2007), were originally developed to measure organic enrichment using macroinvertebrate assemblages. The most frequently observed response to increased nutrient levels in streams is an alteration in species composition and resultant shifts in relative abundance associated with trophic response. The determination of sources, causes, magnitude, and extent of impairments require diagnostic measures that can assess the complexity of water quality impairments resulting from non-point sources of nutrients.

The primary objective of this study is to develop biotic models capable of determining cause and effect between nutrients and fish assemblages. The second objective is to establish an approach for designating defensible nutrient biotic index (NBI) score thresholds and corresponding nutrient concentrations, above which fish assemblages show alterations due to increased nutrient concentrations. By using species occurrence frequency at varying nutrient concentrations as a weighted average, specific nutrient optima can be established that will identify specific nutrient shift response variables (Ter Braak and Juggins 1993; Black et al. 2004). The development of nutrient optima is based on the observation that most species exhibit a unimodal response curve in relation to environmental variables (Jongman et al. 1987). The assignment of tolerance values to species based on individual nutrient parameters would provide the ability to diagnose community data against a linear scale of impairment. The establishment of scoring thresholds to assess impairments would provide quantitative benchmarks for establishing criteria thresholds and enable adaptive management strategy employment for State-wide nutrient criteria development.

2 Methods

2.1 Study Area

This study was conducted to develop nutrient thresholds for Indiana. The study area includes the State of Indiana, which is predominantly cropland with the remaining land uses comprised of pasture, isolated woodlots, and urban areas.

The Corn Belt and Northern Great Plains, Mostly Glaciated Dairy Region and the Southeastern Temperate Forested Plains and Hills Nutrient Ecoregions incorporate all of the State of Indiana (EPA 1998a; Dodds and Oaks 2004). This report analyzes data from the portions of these Nutrient Ecoregions confined by the political boundaries of the State of Indiana hence forth referred to as the Indiana State Wide analysis. This study uses data collected by two designs including a statewide random, probability based sampling approach and an intensive watershed, non-random design to assess stream condition. Random sites were weighted according to stream order so that equal probability of selection across all drainage categories occurred for statewide condition assessment, while the intensive watershed design included spatially intensive sampling to determine cause and effect in smaller hydrologic units (Morris et al. 2006; Simon and Morris 2009).

For the current study, we used Indiana Department of Environmental Management (IDEM) stream sampling results from 1996 to 2007 at 1274 sites (Fig. 1). These streams represent the full range of drainage area sizes and are classified as representative sites (Stoddard et al. 2006), rather than reference sites. They represent a wide variety of natural and anthropogenic conditions. The use of data from the two sampling designs ensures inclusion of a response relationship, capable of capturing both ends of the effect gradient. In this application, a reference site design would limit the utility of the approach since deviation from that established condition would be measured and not necessarily able to pinpoint community shifts occurring at varying resolutions and complexities. Nutrient summary statistics for the Indiana State Wide analysis calculated from spring, summer, and fall ambient water quality monitoring data are included in Table 1. These data were used to calculate the EPA 25th and 75th percentile values based on water quality information for multiple lines of evidence. The calculation of these statistics includes only the random probability data; therefore, the characterization of the nutrient gradients based on these statistics has a high degree of confidence.

2.2 Field Collection

2.2.1 Fish Collection

Daytime, single-pass fish assemblage inventories were conducted using a Smith-Root model 15-D backpack electrofisher, a Smith-Root 2.5 GPP shore/boat electrofisher or a Smith-Root Type VI Boat Electrofishing system from 2000 to 2007 (IDEM 1992). The backpack system was used on small streams with wetted widths < 3.3 m, the 2.5 GPP gear was used on wadeable streams having a wetted width > 3.4 m, and either the 2.5 GPP or the Type VI boat electroshocking systems were used for all other non-wadeable collections. Backpack electrofishing unit settings produced 850 W of power with 300 V output and 2 A, the 2.5 GPP system had 500 V, 3 A, and 2,500 W and the Type

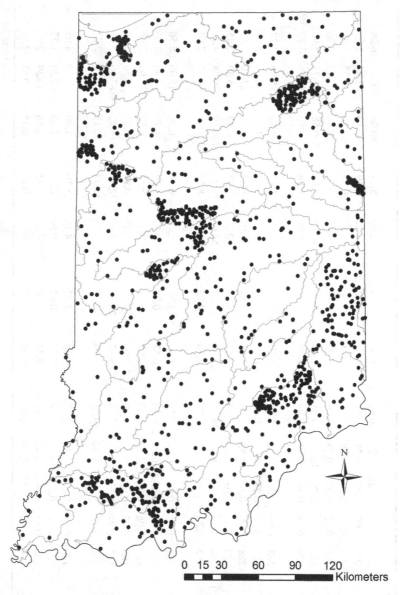

Fig. 1 Map of the State of Indiana showing 1,274 sites sampled for fish assemblages, nutrients and general chemistry from 1996 to 2007

IV 500 V, 3–5 A, and 3,500 W. Sampling time was dependent on habitat complexity, but representative samples were collected within 300–5,000 s. For headwater and wadeable streams, upstream sample distance was 50 m unless stream width exceeded 3.4 m, in which case sample distance was 15 times the wetted width, maximum 500 m. Sampling distances of 11–15 stream widths

Table 1 Probabilistic summary statistics for statewide monitoring surveys in Indiana during 1996–2008 based on log normal distributions

Basin	N	Min	5th	25th	Mean (X)	Median	X + 1/2 SD	75th	X + 1SD	95th	Max	Std. Dev.
All												
NO2 + NO3 (mg/l)	2273	0.01	0.02	0.30	0.82	1.20	2.06	2.95	5.14	10.00	100.00	6.26
TKN (mg/l)	2191	0.05	0.05	0.33	0.57	0.63	0.92	1.02	1.49	2.10	57.00	2.61
Total nitrogen (mg/l)	2102	0.06	0.30	0.99	1.94	2.15	3.48	4.20	6.23	11.48	64.05	3.21
Total phosphorus (mg/l)	2361	0.02	0.02	0.05	0.09	0.09	0.15	0.17	0.25	0.46	7.20	2.76
Chlorophyll a (peri) (mg/m2)	873	0.41	2.79	16.53	37.25	41.43	74.00	88.30	147.01	330.68	1553.74	3.95
Chlorophyll a (phyto) (ug/l)	879	0.03	0.48	1.23	2.94	2.45	5.69	5.69	10.98	40.90	250.89	3.73
West Fork White River												
NO2 + NO3 (mg/l)	283	0.01	0.12	0.52	1.32	1.70	2.73	3.50	5.64	11.00	25.00	4.27
TKN (mg/l)	277	0.05	0.25	0.39	0.66	0.71	0.98	1.10	1.44	1.90	6.00	2.18
Total nitrogen (mg/l)	228	0.06	0.57	1.55	2.73	2.74	4.36	5.07	6.95	13.86	27.10	2.55
Total phosphorus (mg/l)	330	0.02	0.03	0.08	0.14	0.14	0.22	0.29	0.35	0.53	5.30	2.45
Chlorophyll a (peri) (mg/m2)	64	1.00	1.26	6.05	18.40	16.44	38.00	56.04	78.45	180.19	379.38	4.26
Chlorophyll a (phyto) (ug/l)	67	0.22	0.38	1.20	2.73	2.70	4.72	5.87	8.17	11.96	56.10	2.99
East Fork White River												
NO2 + NO3 (mg/l)	162	0.01	0.03	0.30	0.79	1.10	1.87	2.70	4.41	6.50	15.00	5.56
TKN (mg/l)	157	0.05	0.05	0.54	0.63	0.67	0.90	0.87	1.29	1.40	4.40	2.06
Total nitrogen (mg/l)	154	0.06	0.51	1.10	1.93	2.07	3.15	3.71	5.14	6.87	15.05	2.66
Total phosphorus (mg/l)	165	0.02	0.03	0.06	0.09	0.09	0.13	0.13	0.19	0.22	2.50	2.12
Chlorophyll a (peri) (mg/m2)	112	0.94	2.83	18.60	37.44	45.97	70.45	73.62	132.57	216.42	1187.56	3.54
Chlorophyll a (phyto) (ug/l)	112	0.16	0.59	1.82	4.11	4.00	7.42	8.28	13.37	30.13	66.13	3.25
Whitewater												
NO2 + NO3 (mg/l)	140	0.01	0.06	1.80	2.02	2.90	4.07	4.10	8.19	7.80	13.00	4.05
TKN (mg/l)	133	0.05	0.23	0.36	0.46	0.47	0.62	0.60	0.82	1.10	2.50	1.77
Total nitrogen (mg/l)	133	0.35	0.69	2.42	3.14	3.44	4.50	4.66	6.45	8.16	13.39	2.05
Total phosphorus (mg/l)	140	0.02	0.02	0.03	0.05	0.05	0.07	0.08	0.10	0.15	0.53	2.11

(continued)

Table 1 (continued)

Basin	N	Min	5th	25th	Mean (X)	Median	X + 1/2 SD	75th	X + 1SD	95th	Max	Std. Dev.
Chlorophyll a (peri) (mg/m2)	94	1.62	7.52	38.21	59.57	62.97	100.40	110.32	169.21	313.99	416.87	2.84
Chlorophyll a (phyto) (ug/l)	94	0.09	0.80	1.46	2.48	2.68	3.72	3.91	5.57	7.43	24.36	2.24
Ohio River tribs.												
NO2 + NO3 (mg/l)	282	0.01	0.01	0.14	0.38	0.48	0.94	1.50	2.31	3.80	40.00	6.03
TKN (mg/l)	223	0.05	0.05	0.38	0.53	0.58	0.86	0.92	1.38	2.10	7.20	2.61
Total nitrogen (mg/l)	221	0.06	0.06	0.73	1.22	1.39	2.19	2.59	3.94	5.45	41.80	3.23
Total phosphorus (mg/l)	284	0.02	0.02	0.03	0.06	0.06	0.11	0.12	0.18	0.36	3.20	2.78
Chlorophyll a (peri) (mg/m2)	121	1.13	3.07	12.77	24.87	26.43	42.68	54.37	73.25	106.86	332.04	2.94
Chlorophyll a (phyto) (ug/l)	121	0.05	0.35	0.86	2.00	1.50	3.95	3.89	7.78	23.66	41.19	3.89
Great Lakes												
NO2 + NO3 (mg/l)	261	0.05	0.05	0.49	1.19	1.50	2.65	3.40	5.89	15.00	45.00	4.94
TKN (mg/l)	262	0.25	0.25	0.25	0.81	0.90	1.33	1.50	2.19	3.20	57.00	2.71
Total nitrogen (mg/l)	249	0.30	0.30	1.32	2.69	2.65	4.64	5.15	8.01	19.10	57.12	2.98
Total phosphorus (mg/l)	274	0.02	0.03	0.08	0.15	0.13	0.25	0.30	0.44	0.95	7.20	3.01
Chlorophyll a (peri) (mg/m2)	118	2.34	8.45	27.09	52.09	48.19	90.06	99.94	155.71	390.77	1012.61	2.99
Chlorophyll a (phyto) (ug/l)	118	0.41	0.62	1.26	3.52	2.31	7.05	8.73	14.13	64.18	160.53	4.01
Upper Wabash												
NO2 + NO3 (mg/l)	215	0.01	0.16	1.30	1.88	2.30	3.62	4.10	6.95	8.00	14.00	3.69
TKN (mg/l)	213	0.05	0.32	0.58	0.79	0.78	1.10	1.10	1.54	2.00	24.00	1.96
Total nitrogen (mg/l)	212	0.06	1.04	2.05	3.19	3.18	4.66	5.10	6.81	9.40	28.50	2.13
Total phosphorus (mg/l)	216	0.02	0.02	0.06	0.09	0.10	0.15	0.15	0.24	0.40	2.60	2.57
Chlorophyll a (peri) (mg/m2)	111	0.48	2.64	14.86	30.38	40.52	59.12	78.93	115.06	150.70	330.68	3.79
Chlorophyll a (phyto) (ug/l)	111	0.13	0.34	0.80	1.97	2.02	3.53	3.38	6.33	12.53	69.80	3.22
Lower Wabash												
NO2 + NO3 (mg/l)	317	0.01	0.01	0.16	0.60	0.80	1.79	3.30	5.36	11.00	24.20	8.99
TKN (mg/l)	311	0.05	0.05	0.20	0.36	0.39	0.66	0.77	1.21	2.10	5.30	3.33

(continued)

Table 1 (continued)

Basin	N	Min	5th	25th	Mean (X)	Median	X + 1/2 SD	75th	X + 1SD	95th	Max	Std. Dev.
Total Nitrogen (mg/l)	311	0.06	0.06	0.63	1.46	1.78	3.03	4.25	6.29	11.64	24.49	4.32
Total Phosphorus (mg/l)	317	0.02	0.02	0.04	0.08	0.08	0.13	0.14	0.21	0.41	1.60	2.59
Chlorophyll a (peri) (mg/m2)	128	0.41	1.63	9.40	39.53	49.99	98.55	148.91	245.67	689.54	1553.74	6.21
Chlorophyll a (phyto) (ug/l)	129	0.03	0.54	1.10	4.42	2.71	10.80	9.58	26.42	137.91	165.99	5.98
Kankakee												
NO2 + NO3 (mg/l)	331	0.01	0.02	0.33	0.85	1.04	2.18	2.82	5.63	13.20	64.00	6.65
TKN (mg/l)	315	0.05	0.05	0.25	0.46	0.63	0.78	1.00	1.32	1.50	6.15	2.89
Total nitrogen (mg/l)	314	0.06	0.18	0.93	1.77	1.79	3.36	3.68	6.38	12.56	64.05	3.59
Total phosphorus (mg/l)	332	0.02	0.02	0.04	0.08	0.08	0.12	0.14	0.19	0.33	4.70	2.51
Chlorophyll a (peri) (mg/m2)	125	1.45	3.31	18.15	45.36	47.92	90.67	123.92	181.25	394.81	1000.74	4.00
Chlorophyll a (phyto) (ug/l)	127	0.33	0.57	1.47	2.99	2.59	5.09	4.08	8.69	21.50	250.89	2.91

are generally adequate to sample a single habitat cycle (Leopold et al. 1964). Boat electrofishing sites were sampled for 500 m along both banks. Fish identified in the field were vouchered for later taxonomic verification, while all other specimens were preserved in 10% formalin for laboratory identification using standard taxonomic references (Becker 1983; Etnier and Starnes 1993) and verified by a regional taxonomic expert.

2.2.2 Water Chemistry

Grab water chemistry samples were collected 1–3 times per year at each site in 1,000 mL certified contaminant free sample bottles from the visual centroid of flow. Sampling devices were cleaned and then rinsed with de-ionized water after each use and placed in clean storage for transport between sites. Once water samples were taken and preservatives were added (2 ml sulfuric acid for nutrients), the exteriors of all sample bottles were rinsed with de-ionized water and placed in ice filled coolers for transport to the laboratory. Duplicate water samples, Matrix Spike (MS)/Matrix Spike Duplicates (MSDs) and field blanks were collected at a rate of 1 for every 20 samples or 1 sample per week when less than 20 samples were taken. Standard field parameter measurements were taken with either a YSI^{TM} multi-parameter water-chemistry analysis unit or a Hydrolab^{TM} data sonde. Field water quality parameters included pH, temperature ($^{\circ}$C), specific conductance (μS), turbidity (NTU), and dissolved oxygen (ppm). All analytical methods followed those outlined in Indiana Department of Environmental Management, Office of Water Quality, Assessment Branch's Quality Assurance Project Plan (IDEM 2004).

2.2.3 Chlorophyll and Algal Biomass Sampling

Phytoplankton was collected using a plastic 3 L sample bottle. If stream flow was less than 1.5 ft/s, a single grab sample was taken from the center of the established transect (linear transect perpendicular to the shore having a width equal to the stream width). If stream flow was greater than 1.5 ft/s, a vertical composite sample was collected from various depths along the transect (Shelton 1994). All samples were stored out of direct sunlight to prevent degradation of photosynthetic pigments until on-site processing could be done. Chlorophyll a (phytoplankton) was filtered from a 100–250 mL aliquot of the original 3 L sample through a 47 mm glass fiber filter. Expressed filters were folded into quarters, wrapped in aluminum foil, placed in small petri dishes and stored on dry ice for transport back to the laboratory (Moulton et al. 2002).

Periphyton samples were collected from a single substrate type according to the following substrate priority: (1) riffles in shallow streams with coarse-grained substrates (epilithic habitat); (2) woody debris in streams with fine-grained substrates (epidendric habitat); and (3) sandy depositional areas along stream margins (epipsammic habitat). If the primary substrate was not present, the progression continued through the priority list; however, once the substrate priority was established that same substrate was collected for each subsequent sampling event.

Epilithic samples consisted of 5 randomly chosen rocks with visible algae growth. The method of epilithic sampling from 2001 to 2003 was the "top-rock scrape", which consisted of scraping the algae from the rock surface and determining the area with a template (Moulton et al. 2002). From 2004 to present, epilithic substrate was sampled from a predetermined surface area using a calibrated syringe sampling device (Moulton et al. 2002). Periphyton samples were rinsed with tap water into a 500 mL sample bottle and filled to a standard volume, usually 400 mL with tap water (Moulton et al. 2002).

Epidendric samples were collected from five submerged woody substrates from areas within or as close to the transect as possible. Woody substrates measured approximately 7 to 10 cm long and 2 to 4 cm in diameter and had visible algae growth. A hard bristle tooth brush was used to scrub the entire surface of all five woody substrates followed by rinsing with tap water into a plastic tub. The sample from the plastic tub was rinsed into a 500 mL bottle adding additional tap water to ensure the bottle was filled to a standard volume. The length and diameter of each cleaned woody substrate was measured and total sampling area calculated (Moulton et al. 2002).

Epipsammic samples were collected from five locations at each site having a depositional zone consisting of sand or silt substrates. Samples were collected by pressing the top half of a 47 mm petri dish into the substrate and then sliding a spatula beneath it to remove a 88.31 mm^3 standardized slice. Sediment was composited by rinsing the petri dish contents into a 500 mL sample bottle with tap water to a standard 400 mL volume (Moulton et al. 2002).

Filtration procedures for the analysis of Chlorophyll a (periphyton) used the same technique independent of priority substrate type. Approximately a 3 mL aliquot of the priority substrate slurry was filtered by aspiration through a 47 mm glass fiber filter. The filter was then folded into quarters, wrapped in aluminum foil, placed in a small petri dish, and stored on dry ice for delivery to the laboratory (Moulton et al. 2002).

Concentrations of Chlorophyll a were measured using a Turner Designs TD-700 fluorometer outfitted for Chlorophyll a analysis following USEPA method 445 with two exceptions (Arar and Collins 1997). Filters were ground in Nalgene centrifuge tubes rather than glass to counter tube breakage and samples were centrifuged at a slower rate generating approximately 320 to 569 g for 15 min as opposed to 675 g as prescribed by EPA. This modified method has been shown to produce comparable results. Lowe et al. (2008), compared the two methods testing 90 Chlorophyll a (periphyton) replicates and 89 Chlorophyll a (phytoplankton) replicates. They used a Wilcoxon Rank Sum test to evaluate whether the median differences between paired samples was zero. No Statistical differences were detected for either Chlorophyll a (periphyton) ($p = 0.977$) or Chlorophyll a (phytoplankton) ($p = 0.715$).

3 Data Analysis

3.1 Data Censoring

Sampling events occurring from 1996 to 2007 resulted in 1294 sites being available for analysis. Because water chemistry and algal biomass sampling occurred independent of fish collections, and at multiple times throughout the year, it was necessary to use a data censoring tool that would capture only those chemistry and algal biomass sampling events occurring within a 90 day window, prior to fish collections (Smith et al. 2007). If more than one chemistry or algal biomass event occurred within this window, the event occurring nearest to the date including or prior to fish collections was used. After the appropriate nutrient data were assigned to each reach, nutrient concentrations were evaluated for replication of observations along the nutrient concentration gradient. Lack of replication in the high nutrient concentration gradient could potentially drive relationships that are not supported by the body of data. To prevent this situation we evaluated each nutrient range of concentrations to flag for removal of extremely high values sharing minimal, if any, replication. Nutrient concentrations for these reaches were generally greater than two times the standard deviation.

3.2 Drainage Classification

Smith et al. (2007) used a modification of a thermal tolerance model (Brandt 2001) to establish optimal nutrient concentrations for macroinvertebrate species in Idaho. This approach used species occurrence data divided into distinct nutrient concentration ranges to calculate, using a weighted averaging approach, optimal nutrient concentrations (optima) that explain species occurrence patterns. These optima were used to develop nutrient tolerance score data for each species. We modified this approach for fish species occurrences which are influenced by natural mechanisms that could bias the calculation of optima. It is well documented that fish species occurrence is heavily influenced by drainage area relationships, which indicates that any given fish species does not have an equal probability of occurrence across varying drainage areas. In order to consider this factor, we broke our data into three drainage class categories. The first data set (small streams) included all reaches having a drainage area less than 100 sq. mi. (n = 1047). The second data set (medium streams) included all reaches with drainage areas greater than 100 but less than 2300 sq. mi. (n = 211) and the final data set (large streams) included all sites with drainage areas greater than 2300 sq. mi. (n = 36). For the purpose of nutrient optima calculation each of these three data sets were analyzed independently.

Of the 1294 available sites, only 230 had corresponding Chlorophyll data. Thus, fewer sites were available in each drainage class for analysis of these variables. Drainage area class memberships for Chlorophyll sites were; small streams (n = 145), medium streams (n = 72) and large streams (n = 13).

3.3 Jenks Analysis

Outliers were removed twice, once for the entire data set and then again before beginning nutrient optima calculations. The second time outliers were removed in the nutrient data within each drainage class. Next, we separated each drainage class data set into 15 ranges or "bins" using the Jenks optimization method in Arc GIS 9.3 (Jenks 1977). The Jenks optimization method classifies data using natural breaks that minimize the squared deviations of the class means thereby maximizing the goodness of variance fit. Once bin ranges were calculated, each reach was assigned a bin with respect to the nutrient concentration observed at that reach. These bin assignments were used to populate the species occurrence data bin model for nutrient optima calculation following Smith et al. (2007).

3.4 Drainage Category Prevalence Determination

Fish species occur across a wide range of drainages; however, they tend to be most prevalent within specific ranges. For example, Golden Redhorse (*Moxostoma erythrurum*) occurs in all three drainage classes; but their percent occurrence within each class varies. They occur in approximately 14% of sites in the small streams class, 67% of reaches in the medium streams class, and 37% of reaches in the large rivers class.

Smith et al. (2007) removed any taxa that occurred in less than 2% of samples, which was a necessary step to ensure accurate calculation and reduce the propensity of producing a Type I error due to low sample size. We chose to set a more conservative threshold for inclusion. For our three drainage classes we excluded species occurring in less than 5% of sites in the small streams class, 15% in the medium streams class and 25% in the large river class. The difference in percent inclusion across the three drainage classes is relative to the available number of reaches within each class.

Based on these observations, the optimal data set for calculating Golden Redhorse nutrient optima would be from both the medium and large river data sets. Following this approach, all species were reviewed for percent occurrence within drainage class. The most representative drainage class or classes were selected for nutrient optima calculations. This step is critical since the model interprets the absence of a species in relation to nutrient concentration and cannot differentiate whether absences were due to nutrient effect or natural condition.

3.5 Species Optima Calculations

Once drainage class assignments were made, we began the calculation of nutrient optima. Nutrient optima are the weighted mean nutrient concentrations that explain species occurrence patterns. Nutrient optima were calculated by dividing the sum of the weighted proportion of times a species occurred in each bin by the un-weighted proportion of times a species occurred in each bin. For example, if bin 1 represents 100 sites and species A occurs at 50 of those reaches the un-weighted proportion of times species A occurs in bin 1 is 0.50 or 50%. The weighted proportion of times species A occurs in bin 1 is equal to the un-weighted proportion multiplied by the average concentration of the target nutrient for all 100 reaches in bin 1. The final optima value for each species was determined by summing the un-weighted proportions across all bins then dividing by the sum of the weighted proportion.

Once nutrient optima values were calculated, tolerance scores were assigned to each species. Tolerance scores are the final ranking of each species describing the nutrient relationship with fish occurrence along a response gradient. This was done by compiling all the species optima values for each nutrient variable, then dividing the resulting range of optima values into 11 equal parts. Starting with the lowest optima range, tolerance scores (0–10) were assigned. This approach allows the nutrient optima to determine the tolerance scoring rather than simply dividing the number of species into 11 equal parts, which would generate an arbitrary scale. Since some species had membership in multiple drainage classes resulting in multiple tolerance values, these multiple scores were averaged to create a single tolerance score for each species.

Figures 2 and 3 illustrates an example calculation of TN optima and subsequent nutrient tolerance score determination for two darter species, Fantail Darter (*Etheostoma flabellare*) and Orangethroat Darter (*E. spectabile*). Fantail Darter demonstrated a decreasing prevalence relative to TN concentration (Fig. 2), while Orangethroat Darter demonstrated a more ubiquitous or steady prevalence (Fig. 3). Resulting tolerance scores for Fantail Darter and Orangethroat Darter were 0 and 7, respectively.

3.6 Nutrient Biotic Index Calculation

Once tolerance scores were finalized, all sites from the three drainage classes were pooled for calculation of Nutrient Biotic Index (NBI) scores. A biotic index is a univariate biocriterion that explains the relationships between a specific contaminant and the resulting effect in species abundance and tolerance change. Our method follows the approach of Hilsenhoff (1987) where the NBI score is equal to the summation of the number of individuals of a given species multiplied by that species tolerance score divided by the total number of individuals at the site having

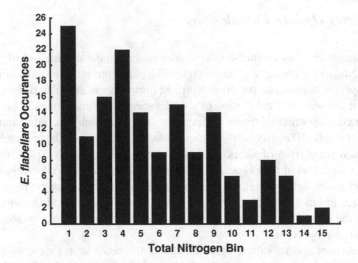

Fig. 2 Bar chart showing the number of sites where Fantail Darters (Etheostoma flabellare) occurred by Total Nitrogen bin assignment. This graph illustrates the decreasing prevalence of Fantail Darter with increasing Total Nitrogen concentration. The final Total Nitrogen tolerance score for Fantail Darter is zero on a scale of 0–10. Zero being the most intolerant and 10 being the most tolerant

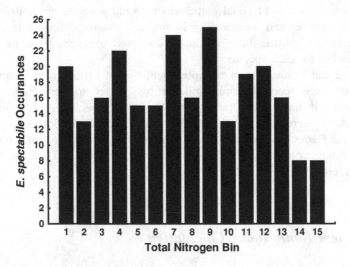

Fig. 3 Bar chart showing the number of sites where Orangethroat Darters (Etheostoma spectabile) occurred by Total Nitrogen bin assignment. This graph illustrates the ubiquitous prevalence of Orangethroat Darter with increasing Total Nitrogen concentration. The final Total Nitrogen tolerance score for Orangethroat Darter is seven on a scale of 0–10. Zero being the most intolerant and 10 being the most tolerant

tolerance scores for each nutrient parameter. Due to our occurrence inclusion rule, not all of the fish species in this study area had sufficient data to determine a tolerance score. Since some NBI scores were calculated from sparse data, i.e., only a small percentage of the total species collected at a site (<5 species) had tolerance scores, we felt it necessary to create a rule limiting the calculation of NBI scores to only those sites having at least six fish species with tolerance values. This step is necessary to minimize the Type I error rate in NBI calculation.

3.7 Statistical Analysis

Once final NBI scores were generated, all sites were divided into 15 bins following the same Jenks procedure described earlier. Relationships between NBI scores and nutrient bin membership and nutrient concentration were determined by generating three dimensional (3D) scatter plots to visualize the interaction so that break points could be determined. A linear/planar fit was applied and variation in slope highlighted via color differentiation (StatSoft 2007). Areas along the curve, which show shifts in NBI score (break points), were further analyzed by generating box plots of NBI scores against nutrient bin assignment to determine statistical significance. Break points identified in the 3D plots were defined in the box plot data and areas between break points labeled Test Response Intervals (TRI). Test Response Intervals are nutrient bins of equal relationship. We compartmentalized TRIs and tested for significant change in NBI score relative to nutrient bin using analysis of variance (ANOVA) ($\alpha = 0.05$). Significant ANOVA models were further analyzed using the Tukey Honest Significant Difference (HSD) post hoc analysis to differentiate between individual TRIs. Test Response Intervals having significantly different mean NBI scores were considered verified as significant and relabeled as Shift Response Intervals (SRI). Non-significant TRI were collapsed and grouped together and presented as a single SRI. Nutrient Biotic Index scores within TRI were evaluated relative to index of biotic integrity (IBI) scores using Spearman Rank Correlation and ANOVA to determine ecological relevance. The Tukey HSD post hoc analysis was used to evaluate mean NBI scores between TRI in relation to IBI integrity class ($\alpha = 0.05$) to verify ecological relevance.

4 Results

4.1 Bin Summary Statistics

Final Jenks bin assignments for each nutrient variable and summary statistics are presented in Tables 2, 3, 4, 5, 6, 7 and 8 and Figs. 4, 5, 6, 7, 8, 9 and 10. Tables 2, 3, 4, 5, 6, 7 and 8 provide upper and lower nutrient concentration

Table 2 Unionized Ammonia bin summary statistics and Jenks Bin assignments

Bin	Bin ranges (mg/L)		n	Bin concentration (mg/L)		Bin NBI		SRI
	Lower limit	Upper limit		Mean ± 95% CI	SD	Mean ± 95% CI	SD	
1	0.00166	0.00215	71	0.00201 ± 0.00003	0.00013	3.3 ± 0.14	0.59	1
2	0.00215	0.00242	151	0.00229 ± 0.00001	0.00008	3.06 ± 0.1	0.65	1
3	0.00242	0.00270	165	0.00256 ± 0.00001	0.00008	2.92 ± 0.09	0.62	1
4	0.00270	0.00304	133	0.00284 ± 0.00002	0.00010	3.03 ± 0.1	0.58	1
5	0.00304	0.00353	83	0.00325 ± 0.00003	0.00013	3.22 ± 0.11	0.50	1
6	0.00353	0.00424	43	0.00383 ± 0.00007	0.00023	3.18 ± 0.14	0.47	1
7	0.00424	0.00534	23	0.00484 ± 0.00012	0.00027	3.08 ± 0.22	0.50	1
8	0.00535	0.00633	25	0.00589 ± 0.00013	0.00031	3.2 ± 0.16	0.38	1
9	0.00633	0.00771	32	0.00701 ± 0.00012	0.00034	3.26 ± 0.21	0.58	1
10	0.00771	0.00991	35	0.00882 ± 0.00019	0.00056	3.11 ± 0.3	0.87	1
11	0.00991	0.01245	30	0.01124 ± 0.00029	0.00079	3.32 ± 0.3	0.80	1
12	0.01245	0.01779	23	0.01435 ± 0.00058	0.00134	3.09 ± 0.3	0.70	1
13	0.01779	0.02939	26	0.02152 ± 0.00123	0.00304	3.33 ± 0.4	0.99	2
14	0.02939	0.06182	9	0.04269 ± 0.00716	0.00931	3.24 ± 0.67	0.88	2
15	0.06182	0.09890	5	0.08514 ± 0.01635	0.01316	3.21 ± 0.77	0.62	2

Summary statistics show the number of sites per bin, mean Unionized Ammonia concentration (mg/L) ± 95% confidence intervals (CI) and the standard deviation. Jenks bin ranges show the maximum and minimum Unionized Ammonia concentration in each bin

Table 3 Nitrogen, Nitrate + Nitrite bin summary statistics and Jenks Bin assignments

Bin	Bin ranges (mg/L)		n	Bin concentration (mg/L)		Bin NBI		SRI
	Lower limit	Upper limit		Mean ± 95% CI	SD	Mean ± 95% CI	SD	
1	0.05	0.113	174	0.058 ± 0.003	0.018	5.58 ± 0.17	1.12	1
2	0.114	0.223	76	0.17 ± 0.008	0.035	5.6 ± 0.24	1.03	1
3	0.224	0.33	68	0.281 ± 0.007	0.028	5.58 ± 0.26	1.06	1
4	0.331	0.46	64	0.393 ± 0.008	0.033	5.58 ± 0.24	0.97	1
5	0.461	0.633	45	0.543 ± 0.015	0.050	5.41 ± 0.33	1.10	2
6	0.634	0.87	61	0.737 ± 0.017	0.067	5.43 ± 0.3	1.16	2
7	0.871	1.17	50	1.008 ± 0.023	0.082	5.14 ± 0.28	0.99	2
8	1.171	1.56	71	1.346 ± 0.024	0.102	5.4 ± 0.22	0.93	2
9	1.561	2.1	86	1.839 ± 0.032	0.149	5.42 ± 0.22	1.02	2
10	2.101	2.72	55	2.468 ± 0.042	0.155	5.85 ± 0.24	0.87	3
11	2.721	3.57	52	3.107 ± 0.066	0.236	6 ± 0.26	0.95	3
12	3.571	4.8	57	4.072 ± 0.092	0.347	5.63 ± 0.29	1.10	3
13	4.801	6.82	44	5.69 ± 0.182	0.598	6.24 ± 0.3	0.97	4
14	6.821	9.8	19	8.189 ± 0.41	0.851	6.29 ± 0.37	0.77	4
15	9.801	13.4	10	11.68 ± 0.842	1.177	6.2 ± 0.84	1.17	4

Summary statistics show the number of sites per bin, mean Nitrogen, Nitrate + Nitrite concentration (mg/L) ± 95% confidence intervals (CI) and the standard deviation. Jenks bin ranges show the maximum and minimum Nitrogen, Nitrate + Nitrite concentration in each bin

Table 4 Total Kjeldahl Nitrogen (TKN) bin summary statistics and Jenks Bin assignments

Bin	Bin ranges (mg/L)		n	Bin concentration (mg/L)		Bin NBI		SRI
	Lower limit	Upper limit		Mean ± 95% CI	SD	Mean ± 95% CI	SD	
1	0.025	0.025	255	0.025 ± -0.025	0.000	2.79 ± 0.14	1.15	1
2	0.026	0.42	18	0.357 ± 0.031	0.062	1.8 ± 0.36	0.73	1
3	0.421	0.54	45	0.503 ± 0.007	0.024	2.73 ± 0.34	1.13	1
4	0.541	0.6	65	0.577 ± 0.005	0.020	3.16 ± 0.31	1.27	2
5	0.601	0.66	41	0.636 ± 0.005	0.016	3.03 ± 0.32	1.03	2
6	0.661	0.72	43	0.695 ± 0.005	0.015	3.04 ± 0.33	1.07	2
7	0.721	0.79	38	0.76 ± 0.006	0.019	3.17 ± 0.36	1.09	2
8	0.791	0.86	48	0.831 ± 0.006	0.020	3.09 ± 0.29	1.00	3
9	0.861	0.95	36	0.909 ± 0.008	0.023	3.51 ± 0.34	0.99	3
10	0.951	1.06	50	0.995 ± 0.006	0.021	3.24 ± 0.27	0.96	3
11	1.061	1.25	42	1.137 ± 0.016	0.051	3.48 ± 0.31	1.01	3
12	1.251	1.51	43	1.379 ± 0.023	0.076	3.56 ± 0.35	1.15	3
13	1.511	1.9	25	1.714 ± 0.042	0.102	2.9 ± 0.45	1.09	3
14	1.901	2.6	25	2.257 ± 0.086	0.208	3.46 ± 0.5	1.22	3
15	2.601	3.4	5	3.066 ± 0.408	0.328	3.11 ± 1.69	1.36	3

Summary statistics show the number of sites per bin, mean TKN concentration (mg/L) ± 95% confidence intervals (CI) and the standard deviation. Jenks bin ranges show the maximum and minimum TKN concentration in each bin

Table 5 Total Nitrogen bin summary statistics and Jenks Bin assignments

Bin	Bin ranges (mg/L)		n	Bin concentration (mg/L)		Bin NBI		SRI
	Lower limit	Upper limit		Mean ± 95% CI	SD	Mean ± 95% CI	SD	
1	0.075	0.225	66	0.108 ± 0.012	0.047	4.5 ± 0.15	0.63	1
2	0.226	0.505	40	0.37 ± 0.023	0.071	4.69 ± 0.24	0.76	1
3	0.506	0.784	65	0.658 ± 0.019	0.075	4.68 ± 0.14	0.57	1
4	0.785	1.05	82	0.921 ± 0.017	0.078	4.64 ± 0.12	0.57	1
5	1.051	1.325	49	1.214 ± 0.023	0.079	4.5 ± 0.12	0.42	1
6	1.326	1.6	44	1.465 ± 0.022	0.071	4.62 ± 0.17	0.57	1
7	1.601	1.88	51	1.752 ± 0.023	0.083	4.84 ± 0.14	0.48	2
8	1.881	2.21	54	2.028 ± 0.024	0.087	4.79 ± 0.14	0.50	2
9	2.211	2.625	55	2.403 ± 0.03	0.110	4.7 ± 0.19	0.70	2
10	2.626	3.17	56	2.872 ± 0.042	0.158	4.87 ± 0.23	0.84	2
11	3.171	3.9	50	3.484 ± 0.059	0.207	4.84 ± 0.19	0.68	2
12	3.901	5.2	52	4.449 ± 0.098	0.353	4.91 ± 0.19	0.69	2
13	5.201	7.11	51	5.986 ± 0.155	0.552	4.84 ± 0.19	0.66	2
14	7.111	9.6	20	8.343 ± 0.306	0.655	5 ± 0.25	0.54	3
15	9.601	13.5	13	12.048 ± 0.718	1.188	5 ± 0.25	0.54	3

Summary statistics show the number of sites per bin, mean Total Nitrogen concentration (mg/L) ± 95% confidence intervals (CI) and the standard deviation. Jenks bin ranges show the maximum and minimum Total Nitrogen concentration in each bin

Table 6 Total Phosphorus bin summary statistics and Jenks Bin assignments

Bin	Bin ranges (mg/L)		n	Bin concentration (mg/L)		Bin NBI		SRI
	Lower limit	Upper limit		Mean ± 95% CI	SD	Mean ± 95% CI	SD	
1	0.025	0.037	57	0.026 ± 0.001	0.002	2.95 ± 0.12	0.46	1
2	0.038	0.054	361	0.05 ± 0	0.001	3.56 ± 0.08	0.77	1
3	0.055	0.073	49	0.065 ± 0.001	0.005	3.16 ± 0.16	0.56	1
4	0.074	0.094	49	0.085 ± 0.002	0.005	3.18 ± 0.15	0.51	1
5	0.095	0.112	74	0.104 ± 0.001	0.005	3.35 ± 0.19	0.81	1
6	0.113	0.134	69	0.125 ± 0.001	0.005	3.49 ± 0.18	0.77	1
7	0.135	0.154	55	0.144 ± 0.001	0.005	3.55 ± 0.22	0.81	1
8	0.155	0.175	49	0.164 ± 0.001	0.005	3.39 ± 0.19	0.67	1
9	0.176	0.2	48	0.187 ± 0.002	0.008	3.45 ± 0.24	0.84	2
10	0.201	0.24	31	0.219 ± 0.004	0.011	3.70 ± 0.31	0.84	2
11	0.241	0.31	37	0.271 ± 0.006	0.018	3.38 ± 0.3	0.90	2
12	0.311	0.4	30	0.354 ± 0.01	0.027	3.61 ± 0.32	0.86	2
13	0.401	0.535	23	0.462 ± 0.018	0.042	3.58 ± 0.34	0.79	2
14	0.536	0.77	9	0.647 ± 0.067	0.087	3.82 ± 0.91	1.19	2

Summary statistics show the number of sites per bin, mean Total Phosphorus concentration (mg/L) ± 95% confidence intervals (CI) and the standard deviation. Jenks bin ranges show the maximum and minimum Total Phosphorus concentration in each bin

Table 7 Chlorophyll a (periphyton) bin summary statistics and Jenks Bin assignments

Bin	Bin ranges (mg/L)		n	Bin concentration (mg/m^2)		Bin NBI		SRI
	Lower limit	Upper limit		Mean ± 95% CI	SD	Mean ± 95% CI	SD	
1	0.41	3.59	10	1.953 ± 0.773	1.080	3.83 ± 0.41	0.58	1
2	3.59	9.97	19	7.105 ± 0.675	1.401	3.58 ± 0.24	0.50	1
3	9.97	15.87	14	14.3 ± 0.703	1.217	3.76 ± 0.31	0.54	1
4	15.87	20.08	10	18.547 ± 0.872	1.219	3.75 ± 0.28	0.39	1
5	20.08	28.38	16	25.729 ± 0.872	1.637	3.88 ± 0.34	0.64	1
6	28.38	34.05	14	31.643 ± 0.959	1.660	4.02 ± 0.31	0.53	2
7	34.05	40.49	13	37.276 ± 1.28	2.118	4.2 ± 0.42	0.70	2
8	40.49	50.09	13	45.93 ± 1.499	2.481	4.09 ± 0.35	0.59	2
9	50.09	61.73	13	56.838 ± 1.731	2.864	3.65 ± 0.48	0.80	2
10	61.73	75.55	23	68.926 ± 1.687	3.901	3.92 ± 0.28	0.64	2
11	75.55	91.56	19	85.014 ± 2.017	4.185	4.25 ± 0.33	0.69	3
12	91.56	113.82	11	103.625 ± 4.488	6.680	4.32 ± 0.37	0.56	3
13	113.82	170.47	15	149.907 ± 7.324	13.226	4.05 ± 0.27	0.48	3
14	170.47	210.30	5	198.937 ± 17.028	13.714	3.98 ± 0.45	0.36	3
15	210.30	663.05	6	518.219 ± 122.59	116.815	4.39 ± 0.59	0.56	3

Summary statistics show the number of sites per bin, mean Chlorophyll a (periphyton) concentration (mg/m^2) ± 95% confidence intervals (CI) and the standard deviation. Jenks bin ranges show the maximum and minimum Chlorophyll a (periphyton) concentration in each bin

limits for inclusion in each bin, along with 95% confidence interval calculations for each bin mean nutrient concentration and NBI scores. Figures 11, 12, 13, 14, 15, 16 and 17 graphically illustrate the relationships of nutrient concentration

Table 8 Chlorophyll a (phytoplankton) bin summary statistics and Jenks Bin assignments

BIN	Bin Ranges (mg/L)		n	Bin Concentration (ug/L)		BIN NBI		SRI
	Lower Limit	Upper Limit		Mean ± 95% CI	SD	Mean ± 95% CI	SD	
1	0.43	0.81	14	0.641 ± 0.072	0.125	3.15 ± 0.4	0.70	1
2	0.81	1.19	15	1.004 ± 0.062	0.112	2.96 ± 0.63	1.14	1
3	1.19	1.62	17	1.453 ± 0.048	0.093	3.63 ± 0.51	0.98	2
4	1.62	2.06	20	1.88 ± 0.055	0.117	3.48 ± 0.37	0.79	2
5	2.06	2.60	13	2.332 ± 0.095	0.158	3.32 ± 0.38	0.63	2
6	2.60	3.28	22	2.97 ± 0.09	0.204	3.35 ± 0.53	1.19	2
7	3.28	4.16	26	3.659 ± 0.105	0.261	3.7 ± 0.43	1.07	2
8	4.16	5.87	17	4.981 ± 0.246	0.479	3.45 ± 0.5	0.97	2
9	5.87	8.25	12	7.347 ± 0.39	0.614	3.67 ± 0.48	0.76	2
10	8.25	10.74	10	9.743 ± 0.395	0.552	3.91 ± 0.85	1.19	3
11	10.74	16.17	9	12.627 ± 1.125	1.464	3.8 ± 0.54	0.71	3
12	16.17	28.64	9	22.529 ± 2.076	2.701	4.63 ± 0.81	1.05	4
13	28.64	41.19	5	36.995 ± 5.091	4.100	4.72 ± 0.85	0.69	4
14	41.19	86.87	5	69.163 ± 14.286	11.505	5.24 ± 0.6	0.49	5
15	86.87	146.23	7	131.663 ± 13.896	15.025	5.58 ± 0.27	0.30	5

Summary statistics show the number of sites per bin, mean Chlorophyll a (phytoplankton) concentration (μg/L) ± 95% confidence values (CV) and the standard deviation. Jenks bin ranges show the maximum and minimum Chlorophyll a (phytoplankton) concentration in each bin

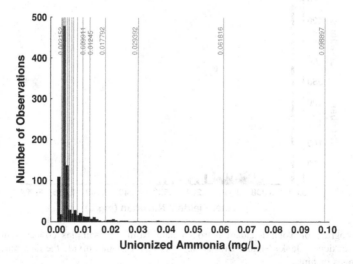

Fig. 4 Histogram of Unionized Ammonia data from 854 samples collected in Indiana. *Vertical lines* denote Jenks Natural Breaks Analysis results which divide the data range into 15 distinct ranges or bins

with Jenks bin assignment. Each nutrient range compartmentalized within bin showed an exponential relationship with the exception of TKN, which was sigmoidal.

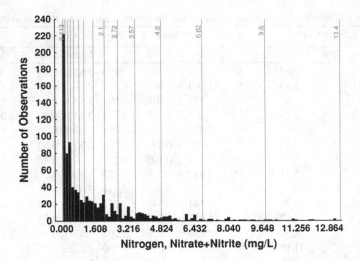

Fig. 5 Histogram of Nitrogen, Nitrate + Nitrite data from 932 samples collected in Indiana. *Vertical lines* denote Jenks Natural Breaks Analysis results which divide the data range into 15 distinct ranges or bins

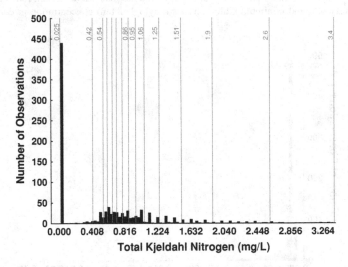

Fig. 6 Histogram of Total Kjeldahl Nitrogen data from 779 samples collected in Indiana. *Vertical lines* denote Jenks Natural Breaks Analysis results which divide the data range into 15 distinct ranges or bins

4.2 Species Optima and Tolerance Classification

We calculated tolerance scores for 77 species of fish based on the nutrient optima calculations (Appendix A).

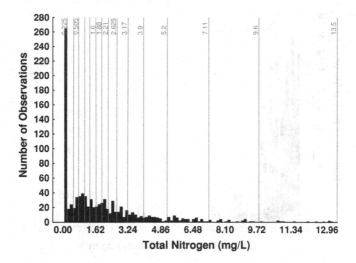

Fig. 7 Histogram of Total Nitrogen data from 748 samples collected in Indiana. *Vertical lines* denote Jenks Natural Breaks Analysis results which divide the data range into 15 distinct ranges or bins

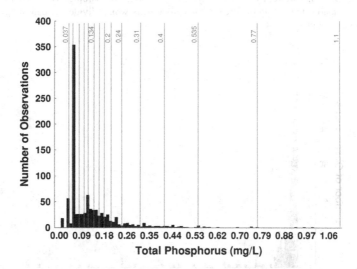

Fig. 8 Histogram of Total Phosphorus data from 941samples collected in Indiana. *Vertical lines* denote Jenks Natural Breaks Analysis results which divide the data range into 15 distinct ranges or bins

4.3 NBI Calculation

Since NBI score calculation is determined only by those species having tolerance scores, we exclude reaches having less than six species having tolerance scores (n = 219). We consider these sites "not assessable" using the NBI since the

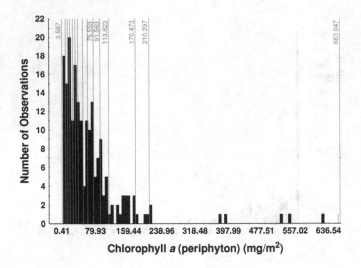

Fig. 9 Histogram of Chlorophyll a (periphyton) data from 201 samples collected in Indiana. *Vertical lines* denote Jenks Natural Breaks Analysis results which divide the data range into 15 distinct ranges or bins

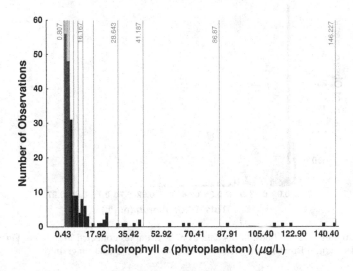

Fig. 10 Histogram of Chlorophyll a (phytoplankton) data from 201 samples collected in Indiana. *Vertical lines* denote Jenks Natural Breaks Analysis results which divide the data range into 15 distinct ranges or bins

calculation using fewer species could increase the models propensity for Type I error; however, this needs to be balanced against requiring too many species there by limiting the overall sample size. Thus, we chose six species as our threshold.

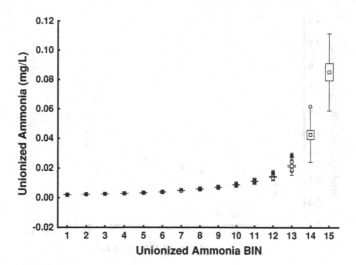

Fig. 11 Box plot of Unionized Ammonia concentrations by 15 bins determined by the Jenks Natural Breaks Analysis. *Boxes* represent the mean and standard deviation while the whiskers represent the standard error of the mean

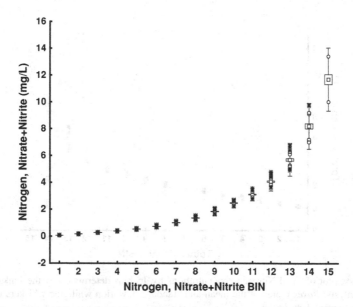

Fig. 12 Box plot of Nitrogen, Nitrate + Nitrite concentrations by 15 bins determined by the Jenks Natural Breaks Analysis. *Boxes* represent the mean and standard deviation while the whiskers represent the standard error of the mean

Additionally, it was necessary to review NBI scores generated from extremely high concentration data. These data were sparse and typically exceeded two standard deviations of the mean (n = 36). Lacking any reasonable replication in

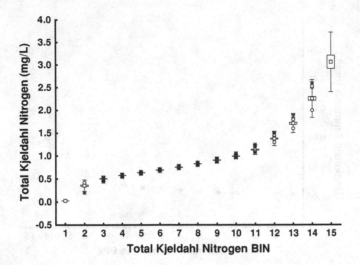

Fig. 13 Box plot of Total Kjeldahl Nitrogen concentrations by 15 bins determined by the Jenks Natural Breaks Analysis. *Boxes* represent the mean and standard deviation while the whiskers represent the standard error of the mean

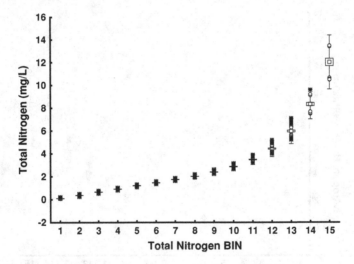

Fig. 14 Box plot of Total Nitrogen concentrations by 15 bins determined by the Jenks Natural Breaks Analysis. *Boxes* represent the mean and standard deviation while the whiskers represent the standard error of the mean

this region of the relationship the precision and accuracy of NBI generated from these sites could not be validated. After all data reductions, the final data set for hypothesis testing was comprised of 963 sites.

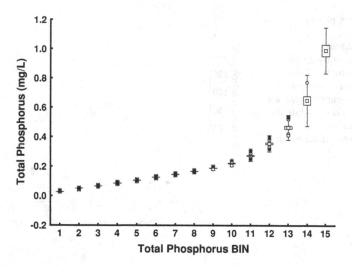

Fig. 15 Box plot of Total Phosphorus concentrations by 15 bins determined by the Jenks Natural Breaks Analysis. *Boxes* represent the mean and standard deviation while the whiskers represent the standard error of the mean

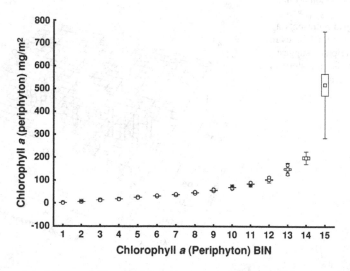

Fig. 16 Box plot of Chlorophyll a (periphyton) concentrations by 15 bins determined by the Jenks Natural Breaks Analysis. *Boxes* represent the mean and standard deviation while the whiskers represent the standard error of the mean

4.4 Test Response Intervals and Shift Response Intervals

Figures 18, 19, 20, 21, 22, 23 and 24 illustrate the 3 dimensional relationships between nutrient concentrations, Jenks bin assignment, and NBI score. Linear/ planar fit functions delineated shift points along the relationship at which the slope

Fig. 17 Box plot of Chlorophyll a (phytoplankton) concentrations by 15 bins determined by the Jenks Natural Breaks Analysis. *Boxes* represent the mean and standard deviation while the whiskers represent the standard error of the mean

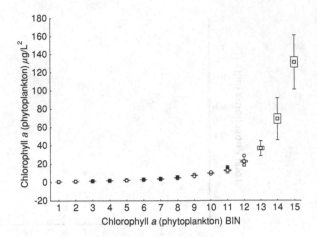

Fig. 18 Three dimensional plot of Unionized Ammonia concentration (mg/L), Unionized Ammonia bin and NBI_Unionized Ammonia score. The *grey* grid represents a linear/planar fit. Changes in color on the plane suggest points at which the slope of the fit changed

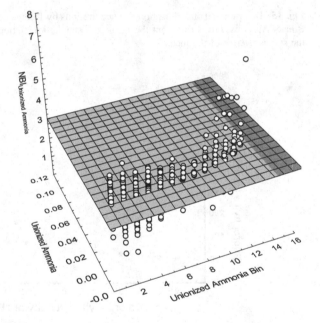

of the fit changes. These relationships are further revealed in Figs. 25, 26, 27, 28, 29, 30 and 31, which show the TRIs. Summary statistics for the TRI chemistry and NBI are shown in Tables 9 and 10, respectively. Figures 32, 33, 34, 35, 36, 37 and 38 show the statistical significance of each TRI response and the resulting SRI based on Tukey HSD post hoc analysis (Table 11). Summary statistics for the SRI based on Tukey HSD post hoc are shown in Tables 12 and 13. Figures 39, 40, 41, 42, 43, 44 and 45 show the relationship between IBI score and NBI scores and Figs. 46, 47, 48, 49, 50, 51 and 52 show the statistical relationships between NBI and IBI integrity classes (Table 14).

Fig. 19 Three dimensional plot of Nitrogen, Nitrate + Nitrite concentration (mg/L), Nitrogen, Nitrate + Nitrite bin and NBI$_{Nitrate+ Nitrite}$ score. The *grey* grid represents a linear/planar fit. Changes in color on the plane suggest points at which the slope of the fit changed

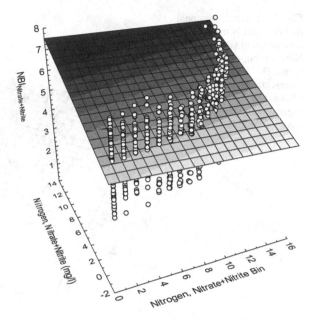

Fig. 20 Three dimensional plot of TKN concentration (mg/L), TKN bin, and NBI$_{TKN}$ score. The *grey* grid represents a linear/planar fit. Changes in color on the plane suggest points at which the slope of the fit changed

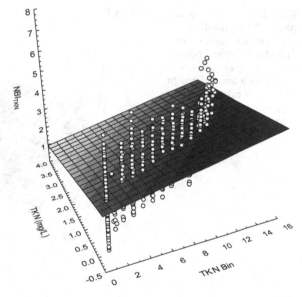

4.4.1 Nitrogen

A single break point was observed for Unionized Ammonia, which was bracketed by two TRIs (Fig. 25). Descriptive Unionized Ammonia concentration and NBI$_{Unionized\ Ammonia}$ statistics for TRI are provided in Tables 9 and 10. The 3D plot showed that a break point occurred between bins 12 and 13 (Fig. 18). This

Fig. 21 Three dimensional plot of Total Nitrogen concentration (mg/L), Total Nitrogen bin and NBI$_{TN}$ score. The *grey* grid represents a linear/planar fit. Changes in color on the plane suggest points at which the slope of the fit changed

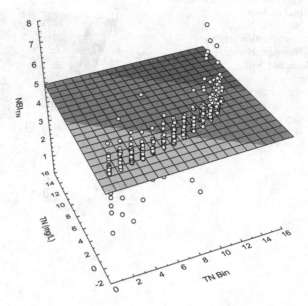

Fig. 22 Three dimensional plot of Total Phosphorus concentration (mg/L), Total Phosphorus bin and NBI$_{TP}$ score. The *grey* gird represents a linear/planar fit. Changes in color on the plane suggest points at which the slope of the fit changed

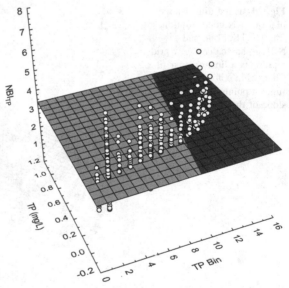

break point was verified by the significant ANOVA model showing an NBI$_{Unionized}$ $_{Ammonia}$ score shift ($p = 0.05$, $\alpha = 0.05$) that occurred between TRI 1 (bins 1–12) and TRI 2 (bins 13–15) (Figs. 25 and 32) resulting in two validated SRIs. The mean concentrations of Unionized Ammonia for SRIs 1 and 2 were 0.003 and 0.03 (mg/L) (Table 12) and the mean NBI$_{Unionized\ Ammonia}$ scores were 3.09 and 3.29, respectively (Table 13). Nutrient Biotic Index$_{Unionized\ Ammonia}$ scores were

Fig. 23 Three dimensional plot of Chlorophyll a (periphyton) concentration (mg/m²), Chlorophyll a (periphyton) bin and NBI$_{Periphyton}$ score. The *grey* grid represents a linear/planar fit. Changes in color on the plane suggest points at which the slope of the fit changed

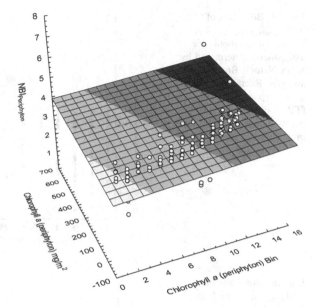

Fig. 24 Three dimensional plot of Chlorophyll a (phytoplankton) concentration (µg/L), Chlorophyll a (phytoplankton) bin and NBI$_{Phytoplankton}$ score. The *grey* grid represents a linear/planar fit. Changes in color on the plane suggest points at which the slope of the fit changed

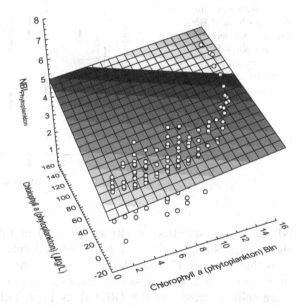

significantly correlated with IBI score (p = 0.00000, α = 0.05) (Fig. 39) and IBI integrity class (p < 0.000, α = 0.05) (Fig. 46). Post hoc analysis showed that NBI$_{Unionized\ Ammonia}$ score statistically predicted two biological integrity ranges (very poor to poor and fair to excellent) (Table 14).

Three break points were observed for Nitrogen, Nitrate + Nitrite among four TRIs (Fig. 26). Descriptive Nitrogen, Nitrate + Nitrite concentration and

Fig. 25 Box plots of
NBI$_{Unionized\ Ammonia}$ by
Unionized Ammonia bin
assignments determined by
Jenks Natural Breaks
analysis. Rectangles identify
Test Response Intervals
(TRI) bracketing the shift
points indicated by the three
dimensional plot

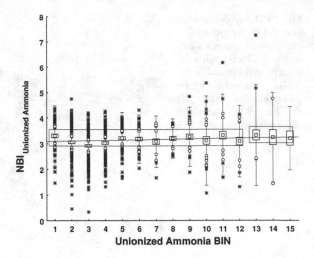

Fig. 26 Box plots of
NBI$_{Nitrate+Nitrite}$ by Nitrogen,
Nitrate + Nitrite bin
assignments determined by
the Jenks Natural Breaks
analysis. *Rectangles* identify
Test Response Intervals
(TRI) bracketing the shift
points indicated by the three
dimensional plot

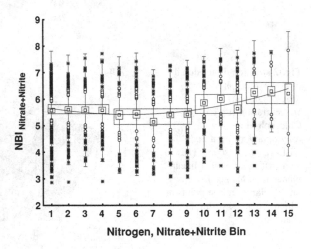

NBI$_{Nitrate+Nitrite}$ statistics for TRI are provided in Tables 9 and 10. The 3D plot
showed that Nitrogen, Nitrate + Nitrite break points occurred between bins 4 and
5, 9 and 10, and 12 and 13 (Fig. 19). The ANOVA model was significant
($p < 0.000$, $\alpha = 0.05$) and post hoc analysis showed a significant NBI$_{Nitrate+Nitrite}$
score shift occurred between TRI 1 (bins 1–4), TRI 2 (bins 5–9), TRI 3 (bins
10–12), and TRI 4 (bins 13–15) (Figs. 26 and 33) resulting in four SRI's having
mean Nitrogen, Nitrate + Nitrite concentrations of 0.13, 1.09, 3.15 and 6.87 mg/
L, respectively (Table 9). The mean NBI$_{Nitrate+Nitrite}$ score for each SRI are 5.58,
5.37, 5.82 and 6.25, respectively (1. 10). Shift Response Intervals 1 and 3 were not
statistically different; however they were separated by SRI 2, which had a significantly lower mean NBI$_{Nitrate+Nitrite}$ score. This relationship produced a convex
curve suggesting an enrichment signature. Nutrient Biotic Index$_{Nitrate+Nitrite}$ scores

Fig. 27 Box plots of NBI$_{TKN}$ by TKN bin assignments determined by the Jenks Natural Breaks analysis. *Rectangles* identify Test Response Intervals (TRI) bracketing the shift points indicated by the three dimensional plot

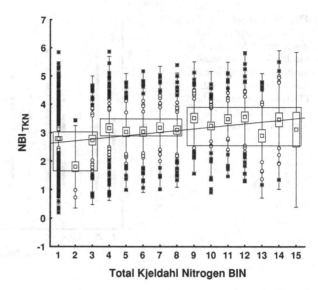

Fig. 28 Box plots of NBI$_{TN}$ by Total Nitrogen bin assignments determined by the Jenks Natural Breaks analysis. *Rectangles* identify Test Response Intervals (TRI) bracketing the shift points indicated by the three dimensional plot

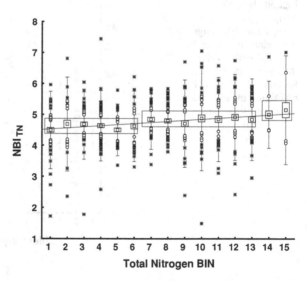

were significantly correlated with IBI score ($p = 0.05$, $\alpha = 0.05$) (Fig. 40) and IBI integrity class ($p = 0.0000$, $\alpha = 0.05$) (Fig. 47). The post hoc analysis showed that the NBI score could significantly predict two integrity ranges (very poor and poor to excellent).

Two break points were observed for TKN (Fig. 27), which were bracketed by three TRIs. Descriptive TKN concentration and NBI$_{TKN}$ statistics for each TRI are provided in Tables 9 and 10. The 3D plot showed that there were two break points occurring between bins 3 and 4 and 8 and 9 (Fig. 20). The ANOVA model was significant ($p < 0.000$, $\alpha = 0.05$) and the post hoc analysis verified mean NBI$_{TKN}$

Fig. 29 Box plots of NBI_{TP} by Total Phosphorus bin assignments determined by the Jenks Natural Breaks analysis. *Rectangles* identify Test Response Intervals (TRI) bracketing the shift points indicated by the three dimensional plot

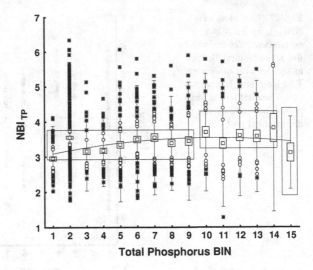

Fig. 30 Box plots of $NBI_{Periphyton}$ by Chlorophyll a (periphyton) bin assignments determined by the Jenks Natural Breaks analysis. *Rectangles* identify Test Response Intervals (TRI) bracketing the shift points indicated by the three dimensional plot

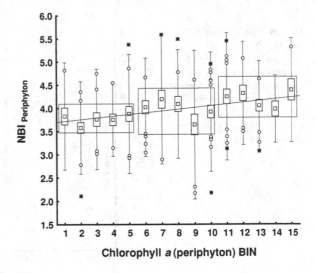

scores were significantly different between TRI 1 (bins 1–3), TRI 2 (bins 4–8) and TRI 3 (bins 9–15) (Figs. 27 and 34), resulting in three valid SRIs. The mean SRI concentrations of TKN were 0.04, 0.68, and 1.27 mg/L (1.12), respectively. The mean NBI_{TKN} scores were 2.73, 3.10, and 3.37, respectively (Table 13). Nutrient Biotic Index$_{TKN}$ scores were significantly related to IBI score (p = 0.00000, α = 0.05) (Fig. 41) and IBI integrity class (p = 0.0000, α = 0.05) (Fig. 48). The post hoc analysis indicated that NBI_{TKN} score could significantly predict three integrity ranges (very poor, poor, and fair to excellent).

Two break points were observed for TN (Fig. 28), which was bracketed by three TRIs. Descriptive TN concentration and NBI_{TN} statistics for each TRI are

Fig. 31 Box plots of NBI$_{Phytoplankton}$ by Chlorophyll a (phytoplankton) bin assignments determined by the Jenks Natural Breaks analysis. *Rectangles* identify Test Response Intervals (TRI) bracketing the shift points indicated by the three dimensional plot

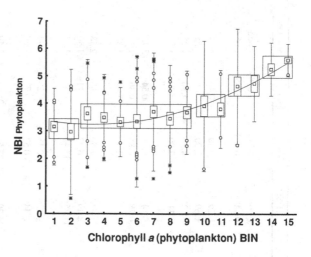

provided in Tables 9 and 10. The 3D plot showed that two break points occurred between bins 6 and 7 and 13 and 14 (Fig. 21). The ANOVA model was significant (p = < 0.000, α = 0.05); however, the post hoc analysis showed that only TRI 1 (bins 1–6) was significantly different resulting in two valid SRIs (bins 1–6 and bins 7–15). The mean SRI concentrations of TN were 0.56 mg/L and 3.30 mg/L (Table 12), respectively. The mean NBI$_{TN}$ scores were 4.60 and 4.85, respectively (Table 13). Nutrient Biotic Index$_{TN}$ scores were not significantly related to IBI score (Fig. 42) or IBI integrity class (Fig. 49).

4.4.2 Phosphorus

Two break points were observed for TP (Fig. 29), which were bracketed by three TRIs. Descriptive TP concentration and NBI$_{TP}$ statistics for each TRI are provided in Tables 9 and 10. The 3D plot showed that the two break points occurred between bins 9 and 10 and 14 and 15 (Fig. 22). The ANOVA model for the three TRIs was not significant. We removed the four reaches from TRI 3 because this TRI was only composed of four data points (the fewest of any TRI in this study). After removal of TRI 3 the ANOVA was significant (p = 0.04, α = 0.05) (Fig. 36) resulting in two valid SRIs (bins 1–9 and bins 10–14). The mean SRI concentrations of TP were 0.07 and 0.32 mg/L (Table 12), respectively. The mean NBI$_{TP}$ scores were 3.43 and 3.58, respectively (Table 13). Nutrient Biotic Index$_{TP}$ scores were significantly related to IBI score (p = 0.00000, α = 0.05) (Fig. 43) and IBI integrity class (p = 0.0000, α = 0.05) (Fig. 50). Post hoc analysis showed that NBI$_{TP}$ scores were not significantly different between very poor and good to excellent class ranges but could significantly predict poor, fair, and good to excellent class ranges.

Table 9 Nutrient concentration summary statistics for Test Response Intervals

Variable	N	Min	5th	25th	Mean (X)	Median	X + 1/2 SD	75th	X + 1 SD	95th	Max	Std. Dev.
Unionized Ammonia (mg/l)												
SRI 1	814	0.002	0.002	0.002	0.003	0.003	0.004	0.004	0.006	0.011	0.018	1.666
SRI 2	40	0.018	0.018	0.020	0.029	0.024	0.038	0.039	0.049	0.091	0.099	1.669
NO2 + NO3 (mg/l)												
SRI 1	382	0.05	0.05	0.05	0.13	0.13	0.19	0.29	0.29	0.40	0.46	2.29
SRI 2	313	0.47	0.51	0.74	1.09	1.19	1.36	1.60	1.69	2.00	2.10	1.55
SRI 3	164	2.20	2.30	2.60	3.15	3.10	3.52	3.80	3.93	4.50	4.80	1.25
SRI 4	73	4.90	4.90	5.40	6.87	6.30	7.91	8.00	9.11	12.00	13.40	1.33
TKN (mg/l)												
SRI 1	318	0.03	0.03	0.03	0.04	0.03	0.08	0.03	0.14	0.51	0.54	3.19
SRI 2	234	0.54	0.55	0.60	0.68	0.68	0.73	0.78	0.78	0.85	0.86	1.15
SRI 3	227	0.86	0.90	1.00	1.27	1.13	1.49	1.50	1.75	2.40	3.40	1.37
Total Nitrogen (mg/l)												
SRI 1	346	0.08	0.08	0.36	0.56	0.80	0.90	1.10	1.44	1.50	1.60	2.58
SRI 2	369	1.62	1.73	2.10	2.99	2.82	3.67	4.10	4.49	6.23	7.11	1.50
SRI 3	33	7.50	7.50	8.03	9.61	8.91	10.62	11.34	11.75	13.43	13.50	1.22
Total Phosphorus (mg/l)												
SRI 1	811	0.03	0.03	0.05	0.07	0.05	0.10	0.12	0.13	0.18	0.20	1.74
SRI 2	130	0.21	0.21	0.25	0.32	0.30	0.38	0.40	0.44	0.57	0.77	1.39
Chlorophyll a (peri) (mg/m2)												
SRI 1	69	0.41	1.70	6.98	10.15	14.19	16.34	20.02	26.30	27.54	28.38	2.59
SRI 2	76	28.87	30.79	36.35	48.46	49.17	56.37	64.54	65.58	72.70	75.55	1.35
SRI 3	56	77.80	79.57	88.17	134.14	112.32	176.76	162.79	232.93	539.58	663.05	1.74
Chlorophyll a (phyto) (ug/l)												
SRI 1	29	0.43	0.45	0.67	0.80	0.86	0.92	0.97	1.06	1.14	1.19	1.33
SRI 2	127	1.30	1.39	1.99	2.98	3.02	3.80	4.08	4.84	7.29	8.25	1.63
SRI 3	19	8.94	8.94	9.72	10.98	10.74	11.86	12.08	12.81	16.17	16.17	1.17
SRI 4	14	18.73	18.73	22.19	26.74	23.15	30.61	33.09	35.04	41.19	41.19	1.31
SRI 5	12	56.10	56.10	69.26	99.90	114.09	119.52	141.03	142.98	146.23	146.23	1.43

Statistics are based on a log normal distribution and are arranged in an increasing rank order

Table 10 Nutrient Biotic Index score summary statistics for Test Response Intervals

Variable	N	Min	5th	25th	Mean (X)	Median	X + 1/2 SD	75th	X + 1 SD	95th	Max	Std. Dev.
$NBI_{Unionized\ Ammonia}$												
TRI 1	814	0.34	1.98	2.76	3.09	3.15	3.4	3.47	3.71	3.98	6.17	0.62
TRI 2	40	1.44	2.37	2.9	3.29	3.11	3.75	3.63	4.2	4.96	7.26	0.91
$NBI_{Nitrate+Nitrite}$												
TRI 1	382	2.86	3.72	4.8	5.58	5.8	6.11	6.49	6.65	6.97	7.72	1.06
TRI 2	313	3.06	3.47	4.71	5.37	5.43	5.89	6.15	6.4	6.97	7.43	1.03
TRI 3	164	2.76	4.11	5.22	5.82	5.86	6.31	6.59	6.81	7.19	7.56	0.99
TRI 4	73	3.48	4.44	5.59	6.25	6.5	6.72	6.93	7.19	7.53	7.83	0.94
NBI_{TKN}												
TRI 1	318	0.2	0.96	1.87	2.73	2.6	3.3	3.64	3.87	4.64	5.84	1.14
TRI 2	234	0.79	1.33	2.25	3.1	3.04	3.65	3.93	4.21	4.82	5.86	1.11
TRI 3	227	0.9	1.56	2.53	3.37	3.51	3.91	4.2	4.44	4.88	5.8	1.07
NBI_{TN}												
TRI 1	346	1.72	3.82	4.31	4.6	4.61	4.9	4.92	5.19	5.4	7.44	0.59
TRI 2	369	1.49	3.79	4.45	4.83	4.83	5.16	5.25	5.49	5.84	7.05	0.66
TRI 3	33	4.08	4.15	4.65	5.06	4.88	5.4	5.3	5.74	6.87	7.01	0.68
NBI_{TP}												
TRI 1	811	1.77	2.44	2.9	3.43	3.3	3.8	3.84	4.18	4.9	6.34	0.75
TRI 2	130	1.28	2.48	2.88	3.58	3.36	4.02	4.18	4.45	5.19	5.79	0.88
TRI 3	4	2.38	2.38	2.77	3.12	3.28	3.38	3.48	3.64	3.55	3.55	0.52
$NBI_{periphyton}$												
TRI 1	69	2.11	2.97	3.38	3.75	3.68	4.02	4.04	4.29	4.75	5.38	0.54
TRI 2	76	2.17	2.81	3.54	3.97	3.99	4.3	4.42	4.63	4.92	5.59	0.66
TRI 3	56	3.07	3.24	3.78	4.2	4.23	4.49	4.61	4.77	5.12	5.45	0.57
$NBI_{phytoplankton}$												
TRI 1	29	0.55	1.85	2.26	3.05	3.12	3.52	3.64	3.99	4.53	4.62	0.94
TRI 2	127	1.28	1.99	2.89	3.52	3.48	4	4.1	4.47	5.1	5.7	0.95
TRI 3	19	1.63	1.63	3.24	3.85	4.05	4.34	4.78	4.82	5.29	5.29	0.96
TRI 4	14	2.49	2.49	4.08	4.66	4.94	5.11	5.34	5.57	5.74	5.74	0.91
TRI 5	12	4.66	4.66	5.15	5.44	5.66	5.64	5.76	5.84	5.82	5.82	0.4

Statistics are arranged in an increasing rank order

Fig. 32 Box plots
illustrating the significant
shift in $NBI_{Unionized\ Ammonia}$
score occurring between TRI
1 and TRI 2 (p = 0.05,
α = 0.05)

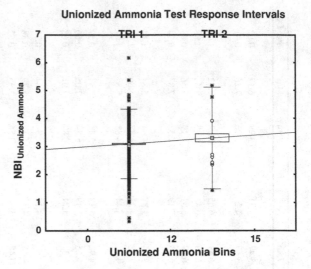

Fig. 33 Box plots
illustrating the significant
relationship between
$NBI_{Nitrate+Nitrite}$ score and
TRI (p < 0.000, α = 0.05).
All TRI have significantly
different mean
$NBI_{Nitrate+Nitrite}$ scores with
the exception of TRI 1 and 3
which are not statistically
different

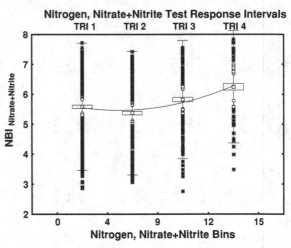

Fig. 34 Box plots
illustrating the significant
relationship between NBI_{TKN}
score and TRI (p < 0.000,
α = 0.05). All TRI have
significantly different mean
NBI_{TKN} scores

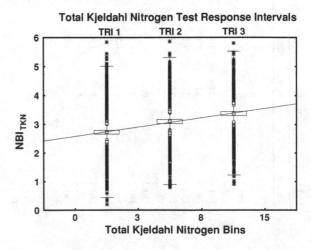

Fig. 35 Box plots illustrating the significant relationship between NBI_{TN} score and TRI ($p < 0.000$, $\alpha = 0.05$). The mean NBI_{TN} score in TRI 1 is significantly different than TRI 2 or 3, while TRI 2 and 3 are not significantly different

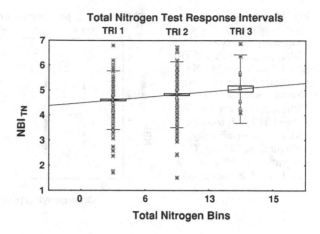

Fig. 36 Box plots illustrating the significant relationship between NBI_{TP} score and TRI ($p < 0.000$, $\alpha = 0.05$)

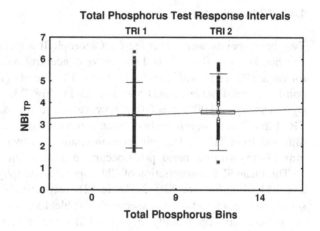

Fig. 37 Box plots illustrating the significant relationship between $NBI_{Periphyton}$ score and TRI ($p < 0.000$, $\alpha = 0.05$). Only the mean $NBI_{Periphyton}$ scores in TRI 1 and TRI 3 are significantly different

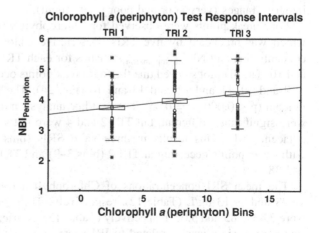

Fig. 38 Box plots illustrating the significant relationship between $NBI_{Phytoplankton}$ score and TRI ($p < 0.000$, $\alpha = 0.05$). No significant differences were observed between mean $NBI_{Phytoplankton}$ scores in TRI 1-2 and TRI 4-5; however, TRI 1-2 are significantly different from TRI 4-5. TRI 3 is significantly different than TRI 1 and 5, but not 2 and 4

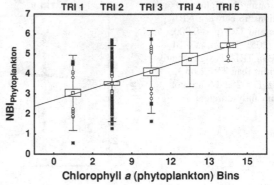

4.4.3 Chlorophyll a

Two break points were observed for Chlorophyll a (periphyton) (Fig. 30), which was bracketed by three TRIs. Descriptive concentration and $NBI_{Periphyton}$ statistics for each TRI are provided in Tables 9 and 10. The 3D plot showed that two break points occurred between bins 5–6 and 10–11 (Fig. 23). The ANOVA model was significant ($p < 0.000$, $\alpha = 0.05$); however, post hoc analysis showed that while TRI 1 and 3 were significantly different from each other, neither were significantly different from TRI 2. This observation resulted in two valid SRIs (bins 1–5 and bins 11–15) with the break point occurring in between bins 6–10.

The mean SRI concentration of Chlorophyll a (periphyton) for SRI 1 was 10.15 and 134.14 mg/m^2 for SRI 2 (Table 12), respectively. The mean $NBI_{Periphyton}$ scores were 3.75 and 4.20, respectively (Table 13). Nutrient Biotic Index$_{Periphyton}$ scores were not significantly related to IBI score (Fig. 44), but were significantly related to IBI integrity class ($p = 0.0023$, $\alpha = 0.05$) (Fig. 51) predicting two integrity ranges (very poor and poor to excellent).

Four break points were observed for Chlorophyll a (phytoplankton) (Fig. 31), which was bracketed by five TRIs. Descriptive Chlorophyll a (phytoplankton) concentration and $NBI_{Phytoplankton}$ statistics for each TRI are provided in Tables 9 and 10. The 3D plot showed that the four break points occurred between bins 2 and 3, 9 and 10, 11 and 12, and 13 and 14 (Fig. 24). The ANOVA model was significant ($p < 0.000$, $\alpha = 0.05$). The post hoc analysis indicated that TRI 1, 3 and 5 were significantly different, but TRI 2 and 4 were not significantly different from adjacent TRIs. This results in three valid SRIs (bins 1–2, 10–12 and 14–15) with break points occurring at TRI 2 (bins 3–9) and TRI 4 (bins 11–13) (Figs. 31 and 38).

The mean SRI concentrations of Chlorophyll a (phytoplankton) were 2.33, 10.98 and 49.13 µg/L (Table 12), respectively. The mean $NBI_{Phytoplankton}$ scores were 3.43, 3.85 and 5.02, respectively (Table 13). Nutrient Biotic Index$_{Phytoplankton}$ scores were significantly related to IBI score ($p = 0.00000$, $\alpha = 0.05$) (Fig. 45)

Table 11 Analysis of variance (ANOVA) and post hoc statistics for NBI Test Response Intervals (TRI)

Variable	ANOVA		Tukey HSD test			
	F	p	TRI 1	TRI 2	TRI 3	TRI 4
$NBI_{Unionized\ Ammonia}$	3.81	**0.05**	–	–	–	–
TRI 2	–	–	**0.05**	–	–	–
$NBI_{Nitrate+Nitrite}$	17.27	**<0.000**	–	–	–	–
TRI 2	–	–	**0.04**	–	–	–
TRI 3	–	–	0.06	**<0.00**	–	–
TRI 4	–	–	**<0.00**	**<0.00**	**0.02**	–
NBI_{TKN}	23.28	**<0.000**	–	–	–	–
TRI 2	–	–	**<0.00**	–	–	–
TRI 3	–	–	**<0.00**	**0.02**	–	–
NBI_{TN}	15.97	**<0.000**	–	–	–	–
TRI 2	–	–	**<0.00**	–	–	–
TRI 3	–	–	**<0.00**	0.11	–	–
$NBI_{TP}\ 3\ TRI$	2.52	0.08	–	–	–	–
TRI 2	–	–	0.09	–	–	–
TRI 3	–	–	0.7	0.47	–	–
$NBI_{TP}\ 2\ TRI$	4.31	**0.04**	–	–	–	–
TRI 2	–	–	**0.04**	–	–	–
$NBI_{Periphyton}$	8.97	**<0.000**	–	–	–	–
TRI 2	–	–	0.67	–	–	–
TRI 3	–	–	**<0.00**	0.67	–	–
$NBI_{Phytoplankton}$	19.17	**<0.000**	–	–	–	–
TRI 2	–	–	0.1	–	–	–
TRI 3	–	–	**0.03**	0.58	–	–
TRI 4	–	–	**<0.00**	**<0.00**	0.1	–
TRI 5	–	–	**<0.00**	**<0.00**	**<0.00**	0.2

Significant p-values are highlighted

and IBI integrity class (p = 0.0000, α = 0.05) (Fig. 52). Post hoc analysis indicated that $NBI_{Phytoplankton}$ scores could significantly predict four integrity ranges (very poor, poor, fair and good to excellent).

5 Discussion

5.1 Reference Condition and Study Design

The EPA approach to nutrient criteria development is based on delineated nutrient ecoregions (EPA 1998b) that have embedded regional reference conditions (Dodds and Oaks 2004). The EPA has suggested three strategies for determining reference condition based on best professional judgement, or using a 75th percentile, or a 25th percentile approach (Buck et al. 2000). The use of a best professional

Table 12 Nutrient concentration summary statistics for significant Shift Response Intervals (SRI)

Variable	N	Min	5th	25th	Mean (X)	Median	X + 1/2 SD	75th	X + 1 SD	95th	Max	Std. Dev.
Unionized Ammonia (mg/L)												
SRI 1	814	0.002	0.002	0.002	0.003	0.003	0.004	0.004	0.006	0.011	0.018	1.666
SRI 2	40	0.018	0.018	0.02	0.029	0.024	0.038	0.039	0.049	0.091	0.099	1.669
Nitrogen, Nitrate + Nitrite (mg/L)												
SRI 1	382	0.05	0.05	0.05	0.13	0.13	0.19	0.29	0.29	0.4	0.46	2.29
SRI 2	313	0.47	0.51	0.74	1.09	1.19	1.36	1.6	1.69	2	2.1	1.55
SRI 3	164	2.2	2.3	2.6	3.15	3.1	3.52	3.8	3.93	4.5	4.8	1.25
SRI 4	73	4.9	4.9	5.4	6.87	6.3	7.91	8	9.11	12	13.4	1.33
Total Kjeldahl Nitrogen (mg/L)												
SRI 1	318	0.03	0.03	0.03	0.04	0.03	0.08	0.03	0.14	0.51	0.54	3.19
SRI 2	234	0.54	0.55	0.6	0.68	0.68	0.73	0.78	0.78	0.85	0.86	1.15
SRI 3	227	0.86	0.9	1	1.27	1.13	1.49	1.5	1.75	2.4	3.4	1.37
Total Nitrogen (mg/L)												
SRI 1	346	0.08	0.08	0.36	0.56	0.8	0.9	1.1	1.44	1.5	1.6	2.58
SRI 2	402	1.62	1.73	2.17	3.3	3.03	4.24	4.59	5.47	8.73	13.5	1.66
Total Phosphorus (mg/L)												
SRI 1	811	0.03	0.03	0.05	0.07	0.05	0.1	0.12	0.13	0.18	0.2	1.74
SRI 2	130	0.21	0.21	0.25	0.32	0.3	0.38	0.4	0.44	0.57	0.77	1.39
Chlorophyll a (periphyton) (mg/m^2)												
SRI 1	69	0.41	1.7	6.98	10.15	14.19	16.34	20.02	26.3	27.54	28.38	2.59
SRI 2	56	77.8	79.57	88.17	134.14	112.32	176.76	162.79	232.93	539.58	663.05	1.74
Chlorophyll a (phytoplankton) (µg/L)												
SRI 1	156	0.43	0.67	1.47	2.33	2.59	3.28	3.71	4.63	7.28	8.25	1.99
SRI 2	19	8.94	8.94	9.72	10.98	10.74	11.86	12.08	12.81	16.17	16.17	1.17
SRI 3	26	18.73	20.77	22.71	49.13	40.82	71.03	111.46	102.69	145.06	146.23	2.09

Statistics are based on a log-normal distribution and arranged in rank order

Table 13 Nutrient Biotic Index (NBI) score summary statistics for significant Shift Response Intervals (SRI)

Variable	N	Min	5th	25th	Mean (X)	Median	X + 1/2 SD	75th	X + 1 SD	95th	Max	Std. Dev.
$NBI_{Unionized\ Ammonia}$												
SRI 1	814	0.34	1.98	2.76	3.09	3.15	3.4	3.47	3.71	3.98	6.17	0.62
SRI 2	40	1.44	2.37	2.9	3.29	3.11	3.75	3.63	4.2	4.96	7.26	0.91
$NBI_{Nitrate+Nitrite}$												
SRI 1	382	2.86	3.72	4.8	5.58	5.8	6.11	6.49	6.65	6.97	7.72	1.06
SRI 2	313	3.06	3.47	4.71	5.37	5.43	5.89	6.15	6.4	6.97	7.43	1.03
SRI 3	164	2.76	4.11	5.22	5.82	5.86	6.31	6.59	6.81	7.19	7.56	0.99
SRI 4	73	3.48	4.44	5.59	6.25	6.5	6.72	6.93	7.19	7.53	7.83	0.94
NBI_{TKN}												
SRI 1	318	0.2	0.96	1.87	2.73	2.6	3.3	3.64	3.87	4.64	5.84	1.14
SRI 2	234	0.79	1.33	2.25	3.1	3.04	3.65	3.93	4.21	4.82	5.86	1.11
SRI 3	227	0.9	1.56	2.53	3.37	3.51	3.91	4.2	4.44	4.88	5.8	1.07
NBI_{TN}												
SRI 1	346	1.72	3.82	4.31	4.6	4.61	4.9	4.92	5.19	5.4	7.44	0.59
SRI 2	402	1.49	3.8	4.48	4.85	4.83	5.18	5.27	5.51	5.85	7.05	0.66
NBI_{TP}												
SRI 1	811	1.77	2.44	2.9	3.43	3.3	3.8	3.84	4.18	4.9	6.34	0.75
SRI 2	130	1.28	2.48	2.88	3.58	3.36	4.02	4.18	4.45	5.19	5.79	0.88
$NBI_{Periphyton}$												
SRI 1	69	2.11	2.97	3.38	3.75	3.68	4.02	4.04	4.29	4.75	5.38	0.54
SRI 2	56	3.07	3.24	3.78	4.2	4.23	4.49	4.61	4.77	5.12	5.45	0.57
$NBI_{Phytoplankton}$												
SRI 1	156	0.55	1.9	2.8	3.43	3.45	3.91	4.04	4.4	5.05	5.7	0.96
SRI 2	19	1.63	1.63	3.24	3.85	4.05	4.34	4.78	4.82	5.29	5.29	0.96
SRI 3	26	2.49	3.43	4.69	5.02	5.26	5.42	5.66	5.83	5.78	5.82	0.81

Statistics are based on a log-normal distribution and arranged in rank order

Fig. 39 Box plot illustrating
the significant relationship
between NBI$_{Unionized\ Ammonia}$
score and IBI score
($p = 0.00000$, $\alpha = 0.05$)

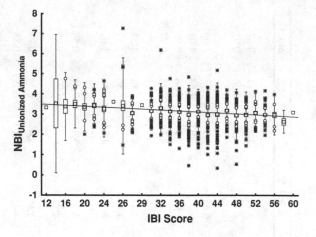

Fig. 40 Box plot illustrating
the significant relationship
between NBI$_{Nitrate+Nitrite}$
score and IBI score
($p = 0.05$, $\alpha = 0.05$)

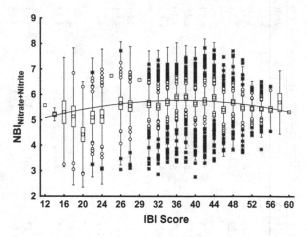

Fig. 41 Box plot illustrating
the significant relationship
between NBI$_{TKN}$ score and
IBI score ($p = 0.0000$,
$\alpha = 0.05$)

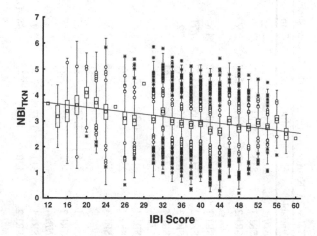

Fig. 42 Box plot illustrating the non-significant relationship between NBI_{TN} score and IBI score

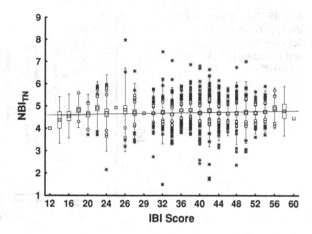

Fig. 43 Box plot illustrating the significant relationship between NBI_{TP} score and IBI score ($p = 0.00000$, $\alpha = 0.05$)

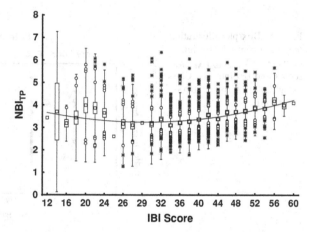

Fig. 44 Box plot illustrating the non-significant relationship between $NBI_{Periphyton}$ score and IBI score

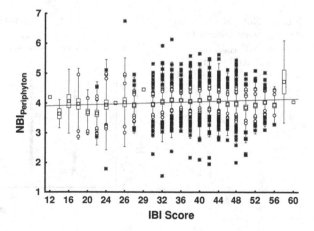

Fig. 45 Box plot illustrating the significant relationship between NBI$_{Phytoplankton}$ score and IBI score (p = 0.00000, α = 0.05)

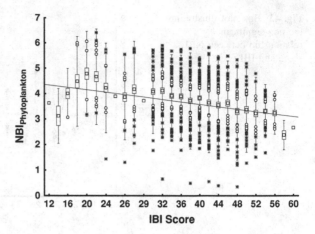

Fig. 46 Box plot illustrating the significant relationship between NBI$_{Unionized\ Ammonia}$ scores and IBI Integrity Class using ANOVA (p = 0.0000, α = 0.05)

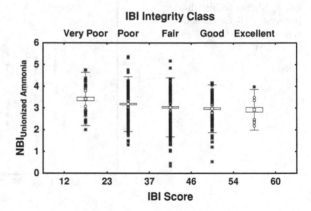

Fig. 47 Box plot illustrating the significant relationship between NBI$_{Nitrate+Nitrite}$ scores and IBI Integrity Class using ANOVA (p = 0.0000, α = 0.05)

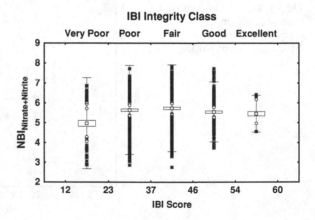

Fig. 48 Box plot illustrating
the significant relationship
between NBI_{TKN} scores and
IBI Integrity Class using
ANOVA (p = 0.0000,
α = 0.05)

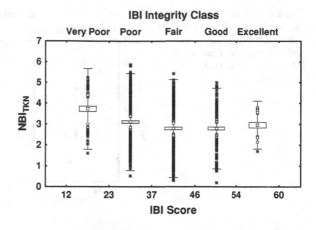

Fig. 49 Box plot illustrating
the non-significant
relationship between NBI_{TN}
scores and IBI Integrity Class
using ANOVA (p > 0.05,
α = 0.05)

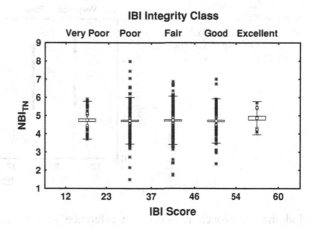

Fig. 50 Box plot illustrating
the significant relationship
between NBI_{TP} scores and
IBI Integrity Class using
ANOVA (p = 0.0000,
α = 0.05)

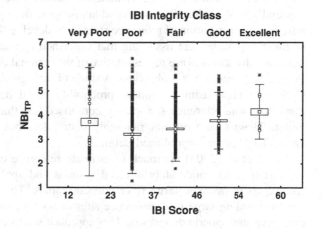

Fig. 51 Box plot illustrating
the significant relationship
between NBI_Periphyton scores
and IBI Integrity Class using
ANOVA (p = 0.0023,
α = 0.05)

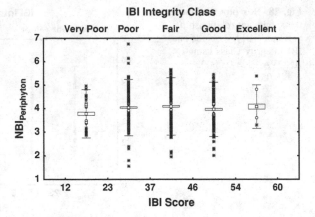

Fig. 52 Box plot illustrating
the significant relationship
between NBI_Phytoplankton
scores and IBI Integrity Class
using ANOVA (p = 0.0000,
α = 0.05)

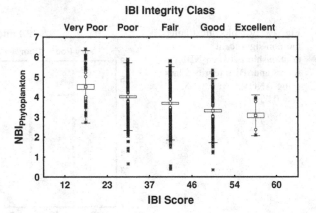

judgment approach requires that reference sites be available; however, in the
absence of reference sites the method fails to accurately predict nutrient values
especially in agriculturally dominated landscapes. Both percentile approaches use
a frequency distribution of reference streams to develop the criteria; however, both
of these approaches are assuming that watershed characteristics show little variation and that samples are representative of the regional condition. In the Corn Belt
Plain this is generally not observed. A study of biological conditions in the Eastern
Corn Belt Plain using a random probability based design found that streams
consistent with reference site quality comprised less than 5% of the entire popu-
lation. These sites were not randomly distributed, but clumped within specific
areas (T.P. Simon, unpublished data).

Smith et al. (2003) approach to estimate reference conditions involved using
data from small, moderately affected systems and applying statistical modeling
techniques to estimate large river nutrient values. This approach eliminates the
need for finding large river reference sites to account for nutrient processing and
changing atmospheric deposition. This approach works only if model assumptions
are met and if small, moderately affected streams are present.

Table 14 Analysis of variance (ANOVA) and post hoc statistics for NBI scores response to IBI integrity class

Variable	ANOVA		Tukey HSD test			
	F	p	Very Poor	Poor	Fair	Good
$NBI_{Unionized\ Ammonia}$	8.3	**<0.000**	–	–	–	–
Poor	–	–	0.07	–	–	–
Fair	–	–	**<0.000**	**0.01**	–	–
Good	–	–	**<0.000**	**<0.000**	0.84	–
Excellent	–	–	**0.03**	0.41	0.95	0.99
$NBI_{Nitrate+Nitrite}$	6.98	**<0.000**	–	–	–	–
Poor	–	–	**<0.000**	–	–	–
Fair	–	–	**<0.000**	0.79	–	–
Good	–	–	**<0.000**	0.83	0.27	–
Excellent	–	–	0.4	0.94	0.79	0.99
NBI_{TKN}	11.37	**<0.000**	–	–	–	–
Poor	–	–	**<0.000**	–	–	–
Fair	–	–	**<0.000**	**<0.000**	–	–
Good	–	–	**<0.000**	**0.02**	1	–
Excellent	–	–	0.06	0.98	0.97	0.97
NBI_{TN}	0.32	0.867	–	–	–	–
Poor	–	–	1	–	–	–
Fair	–	–	1	1	–	–
Good	–	–	0.99	1	0.99	–
Excellent	–	–	0.96	0.86	0.91	0.84
NBI_{TP}	22.87	**<0.000**	–	–	–	–
Poor	–	–	**<0.000**	–	–	–
Fair	–	–	**0.04**	**0.001**	–	–
Good	–	–	0.99	**<0.000**	**<0.000**	–
Excellent	–	–	0.2	**<0.000**	**<0.000**	0.22
$NBI_{Periphyton}$	4.18	**0.002**	–	–	–	–
Poor	–	–	**0.02**	–	–	–
Fair	–	–	**0.002**	0.79	–	–
Good	–	–	0.24	0.61	0.12	–
Excellent	–	–	0.32	1	1	0.93
$NBI_{Phytoplankton}$	34.51	**<0.000**	–	–	–	–
Poor	–	–	**<0.000**	–	–	–
Fair	–	–	**<0.000**	**<0.000**	–	–
Good	–	–	**<0.000**	**<0.000**	**<0.000**	–
Excellent	–	–	**<0.000**	**<0.000**	**0.02**	0.8

Significant results are highlighted

Dodds and Oaks (2004) used a series of land use models to generate reference nutrient concentrations using linear regression. Dodds and Oaks (2004) found that the regression method in the absence of anthropogenic land uses provided comparable values to Smith et al. (2003) and the EPA 25th percentile approach. The regression approach's primary limitation was not quantifying all sources of human impacts due to a lack of information. Non-normal data distribution

Table 15 Total Phosphorus (TP), Total Nitrogen (TN), and Chlorophyll a (periphyton)values for the State of Indiana using chemical probabilistic and nutrient break point values compared to reported reference approaches of Dodds and Oaks (2004), Smith et al. (2003), and EPA's 25th percentile

	TP (μ/L)	TN (μ/L)	Chl. a (phyto) (μ/L)
Corn Belt and Northern Great Plains	–	–	–
Dodds and Oaks (2004)	23	566	–
EPA 25th percentile approach	76	2180	–
Smith et al. (2003)	54	355	–
Mostly Glaciated Dairy Region	–	–	–
Dodds and Oaks (2004)	23	565	–
EPA 25th percentile approach	33	540	–
Smith et al. (2003)	22	147	–
Southeastern Temperate Forested Plains and Hills	–	–	–
Dodds and Oaks (2004)	31	370	–
EPA 25th percentile approach	37	690	–
Smith et al. (2003)	48	150	–
Indiana Approach	–	–	–
IDEM 25th percentile	50	990	1.23
IDEM 75th percentile	170	4200	5.69
Mean Protection Value	70	560	2.33
Significant Biological Effect	–	–	–
NBI	3.34	4.6	3.43
IBI response	42	25[a]	44

[a] non significant value

confounds the relationships since the transformation step to correct for non-normal proportional data is not defined at zero. This requires extrapolation beyond known data points and can induce unpredictable error. The regression approach does not require data from a large number of reference sites or low impact sites. If there are no low impact sites, the method requires prediction of data that is estimated far from the intercept point. Dodds and Oaks (2004) suggest that the greatest accuracy using the regression approach is based on inclusion of sites reflecting a relative continuum of land-use intensity.

Nutrient criteria based on the EPA 25th and 75th percentile approaches are compared to Indiana State Wide probability based values, as well as, Dodds and Oaks (2004) values for the Corn Belt and Northern Great Plains, Mostly Glaciated Dairy Region and the Southeastern Temperate Forested Plains and Hills Nutrient Ecoregions (Table 15). Values for TP and TN showed higher nutrient values associated with Indiana probabilistic observed values than those reported by Dodds and Oaks (2004) or the 25th percentile EPA approach. The Corn Belt Plain loadings of TP and TN are among the highest in the Mississippi River basin (Alexander et al. 2008) so our estimates of higher concentrations based on the probability based design are within the expected range.

5.2 Sample Size Needed to Capture Regional Variation

Indiana Department of Environmental Management uses a probabilistic sample design strategy to conduct their monitoring over a five-year rotation of watersheds. These values represent over 1,200 sites that include multiple visits to each site. This estimate provides a robust estimation of seasonal, annual, and regional variation. Dodds and Oaks (2004) based their regression approach on 42 sites in the six states representing the Corn Belt and Northern Great Plains ecoregions. Based on our sample statistics and variance needed to obtain a 95% confidence interval, we determined that 71 samples were required to capture the variation for total phosphorous, while 1,239 observations were needed for TN. The probabilistic sampling approach is not equivalent to a reference condition approach; however, the sample statistics are representative of the region and show similar concentration benchmarks as Smith et al. (2003).

5.3 Validation of Calibration Procedure

Our primary assumption for the nutrient biotic indices calibration was that the range of nutrient concentrations observed across our dataset was a representative dosing gradient for the Corn Belt Plain. The dosing gradient along with fish species response data would reliably predict species specific optimal nutrient concentrations. These "optima" could then be interpreted into a linear species scale of nutrient tolerance values. We applied the nutrient optima model and generated fish species nutrient tolerance scores that assessed site specific fish assemblage data using a univariate biotic index approach. We wanted to ascertain the confidence in the responsiveness of the NBI to nutrient concentration by testing the null hypothesis that mean NBI score would not be significantly different at one or more discrete points along the dosing gradient (i.e., break points). We predicted break points using a three dimensional model approach and tested significance using ANOVA and Tukey HSD Post Hoc Analysis.

Since our model and its subsequent interpretations were based upon unverified assumptions, we choose to test the application and interpretation of the model using Unionized Ammonia data. Unionized Ammonia is a well known nutrient related contaminant that has an established acute criterion of 8.4 mg N/L (pH adjusted to 8.0 and salmonids excluded), while the chronic criterion is 1.24 mg N/L (pH adjusted to 8.0 at 25°C) (EPA 1999). This step is necessary to both validate the model as well as calibrate interpretation.

Our Unionized Ammonia model predicted a significant alteration in biological structure for Unionized Ammonia concentration at one break point defined by two SRIs (Fig. 25 and 32). The mean concentration of Unionized Ammonia representing SRI 2 was 0.03 mg N/L (Table 12). Thus, we believe that our model is valid, since our predicted Unionized Ammonia concentration eliciting a biological response is consistent with the existing Unionized Ammonia criterion of 0.03 mg/L. Our model approach is consistent with a Criterion Continuous Concentration (CCC) based on long term chronic exposure. Augspurger et al. (2003) reported that acute values for freshwater unionid mussels ranged from 0.3 to 1.0 mg N/L at pH 8.0 and 25°C. Our value is less than the chronic value for unionid mussels, which was recalculated as the CCC of 1.24 mg N/L at pH 8.0 (Augspurger et al. 2003). There are several possible explanations for the difference in Unionized Ammonia criteria values and our results. First, the EPA approach recognizes some allowance for mortality while our approach does not. Second, the EPA approach includes only data for genera that had life cycle toxicity tests while our result takes into consideration the entire fish assemblage and the differing sensitivities. Lastly, our result is based on the multi-species dynamics involved in a web based niche effect which is clearly not considered in the toxicity based approach.

Under normal conditions, Ammonia is converted to Nitrite by microorganisms. Nitrite is toxic at low levels; however, in naturally oxygenated water systems, Nitrite is rapidly oxidized to Nitrate which can then be assimilated by plants. Under normal circumstances, Nitrates only become toxic when conditions favor the reduction of Nitrate to Nitrite. At high Nitrate levels reduction can metabolically occur back to nitrite, which can react with hemoglobin to produce methemoglobin. Methemoglobin limits the system's ability to properly transport oxygen through the blood. In humans, this conversion of hemoglobin to methemoglobin or methemeglobinemia is of primary concern for infants under 3 months of age. Specific NBIs for each Nitrogen component were developed because of the complex cycling of Nitrogen in the environment, varying levels of toxicity associated with each constituent and the resulting affects on structuring biological assemblages.

5.4 Nitrogen Relationships with Fish Assemblages

Our TN model predicted two break points that correspond with mean NBI_{TN} score change; however, post hoc analysis determined that only one break point was valid. The concentration of Nitrogen, Nitrate + Nitrite can be significantly linked to a biological assemblage shift above a threshold value. Provided that Nitrate toxicity is considered to be of negligible risk in flowing waters, the actual contribution of nitrite to this relationship should be considered. If Nitrate is non-toxic and constitutes the vast majority of the combined value, it is logical to assume that either the small contribution of nitrite is responsible for explaining the biological shift or nitrite represents a higher proportion of the total than previously considered. Under certain conditions, Nitrate can be reduced back to nitrite; and, there is evidence to suggest that this process can be mitigated anthropogenically. Our study suggests that the prevalence of nitrite in surface waters should be considered further as a potential chemical response signature.

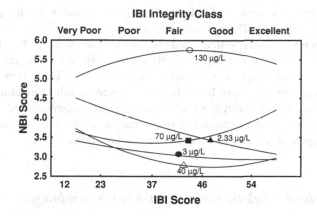

Fig. 53 Relationships between NBI score, IBI score and IBI Integrity Class for each of the five nutrient parameters showing a significant relationship with IBI. The symbol denotes concentration and position along the curve where a significant response was detected. ○ = Nitrogen, Nitrate + Nitrite, = Chlorophyll a (phytoplankton), ■ = Total Phosphorus, ● = Unionized Ammonia, and △ = Total Kjeldahl Nitrogen

The remaining component of TN, which is TKN, represents the organic and Ammonia forms of Total Nitrogen contribution. Total Kjeldahl Nitrogen in natural systems represents the fraction of TN that could, under reducing conditions, contribute to the presence of Unionized Ammonia. Our model predicted two break points for mean NBI_{TKN} score shifts (Fig. 34). These breaks were validated in the post hoc analysis showing that NBI_{TKN} score increased significantly along the TKN concentration gradient (Table 11).

We calibrated four separate NBI models for the Nitrogen cycle. These models show statistically significant break points for biological structure alteration. To test whether this alteration of biological structure was significant and responsible for altering biological integrity of fish assemblages, we tested each NBI for significant interaction with the Index of Biotic Integrity. Unionized Ammonia; Nitrogen, Nitrate + Nitrite; and TKN NBI scores were significantly correlated with IBI score despite varying directional trajectory relationship changes. Nutrient Biotic $Index_{Unionized\ Ammonia}$ scores elicited a classic dose response relationship with IBI confirming a linear relationship between lower $NBI_{Unionized\ Ammonia}$ scores and higher IBI scores as did TKN with higher variation. Nutrient Biotic $Index_{Nitrate+Nitrite}$ showed a significant relationship with IBI score; however, this polynomial function relationship suggests the potential for a stimulus effect in the middle range of IBI scores.

To further understand NBI and IBI score relationships, we tested the effect NBI score had on IBI integrity class (Fig. 53). The response of IBI to both NBI_{TKN} and $NBI_{Nitrate+Nitrite}$ were conflicting. Both produced polynomial relationships; however, NBI_{TKN} demonstrated a concave polynomial relationship while $NBI_{Nitrate+Nitrite}$ had a convex polynomial relationship. The response of NBI_{TN}

(dependent upon the sum of TKN and Nitrogen, Nitrate + Nitrite generates a straight line (Fig. 42). These results demonstrate the complex interaction of Nitrogen species in natural systems and shed some light on why others have found it so difficult to establish significant relationships between TN and biological integrity. Our results suggest that establishing criteria for TN may not be a reasonable management objective for Nitrogen. Apparently, fish assemblages respond to various aspects of the Nitrogen cycle; however, the magnitude and direction of these effects could produce conflicting or inaccurate results when considering only a single component, or a summation of components of the cycle.

5.5 Phosphorus Relationships with Fish Assemblages

Our initial model for evaluating TP effects on fish assemblages was not statistically significant, but was determined to be significant upon further analysis. During the initial stages of data manipulation, we evaluated each nutrient for replication along the entire range of conditions. This was done to limit the effects of sparse high values from significantly affecting a relationship that may not be supported. The effect of insufficient replication in the highest data range caused a non-significant interaction in our model. The Jenks analysis procedure for TP populated bin 15 membership (the highest nutrient concentrations) with only four reaches. The bin mean NBI_{TP} was much lower than those bins preceding it and stood out as potentially spurious (Fig. 29). Once this bin was removed the model was significant and a single break point was validated representing an overall shift in fish assemblage. The NBI_{TP} demonstrated one of the strongest relationships with IBI score resulting in a concave polynomial regression. The direction of the relationship showed that NBI_{TP} score increases with IBI score. This is a counterintuitive result since the NBI_{TP} scale suggests that higher scores denote a negative response in the index; however, our study shows that the more TP the better the biological integrity. The possible mechanism is caused by the bioavailability of Phosphorus. Our TP measure is composed of both dissolved and particulate Phosphorus in the water and does not represent the component that is included in multicellular organisms or the sediment or interstitial water. These fractions are not available and do not represent the saturation point or buffering capacity of the system. The Phosphorus loading under natural conditions should remain in balance with the system's ability for uptake and storage in metabolic processes. When alterations of these natural balances occur, the bioavailable portions of TP exceed the streams natural capacity for assimilation, which can result in algal blooms. The natural process for mitigating Phosphorus removal (i.e., wetland drainage and riparian buffer removal) has been altered through anthropogenic disturbance. Wetlands adjacent to flowing waters increase the assimilative capacity of that system while intact riparian corridors limit the loading of Phosphorus to the system. These limitations in natural assimilation capacity, compounded with increased anthropogenic loading of TP have resulted in increased algal biomass across the United States.

Without the separation of the bioavailable fraction of TP in our study, we believe that our NBI$_{TP}$ may not be measuring the negative effects of dissolved phosphorus, but rather indirectly the effects of system capacity to assimilate Phosphorus. Our hypothesis is that a strong relationship should be observed between habitat quality (factors capable of mitigating Phosphorus availability) and biological integrity. Potential factors confounding this relationship can be eliminated by measurement of the various constituents of Phosphorus rather than only measuring TP.

5.6 Chlorophyll a Relationships with Fish Assemblages

EPA has recommended the use of Chlorophyll a as an early response indicator in lieu of Phosphorus and Nitrogen constituent data. Sampling Chlorophyll a (periphyton) and Chlorophyll a (phytoplankton) as indicators of excess Phosphorus loading was determined through the response of fish assemblages and the development of NBIs. Our calibrated model established that fish assemblages change with increased concentrations of both Chlorophyll a types; however, the effect on biological integrity was variable. The NBI$_{periphyton}$ was not significantly correlated with IBI score or integrity class, while the NBI$_{phytoplankton}$ was strongly correlated with both IBI score and integrity class.

5.7 Numerical Nutrient Criteria Derivation

The purpose of the NBI model is to protect the assemblage structure prior to the SRI. The observed break point occurs as a response to assemblage alteration, thus the SRIs are statistically observed changes. In order to prevent assemblage alteration, the criteria derivation should protect for nutrient concentrations that are conservative. Thus, the mean of SRI prior to the break point is protective against the observed shift. We chose the first SRI in order to be most conservative. The numerical criteria can be set in numerous points associated with the various SRI changes; however, we chose to establish our numerical nutrient criteria based on the mean of the SRI prior to the first observed shift.

The realized consequence of the shift is noted in the comparison of the concentration to the IBI. Even though SRIs can be validated, significant response with the IBI or integrity condition were not always able to be demonstrated. For example, two break points were validated for TN showing a significant relationship with NBI$_{TN}$; however, the NBI$_{TN}$ elicits a non-significant response with the biological condition. So even though significance with a biological assemblage response can be obtained, it may not necessarily correspond to a measureable biological condition. Managing for TN will not provide sufficient clarity for demonstrating the needed result in Indiana, but may be protective for loadings into the Gulf of Mexico. A more relevant measure of the Nitrogen cycle in Indiana should be directed towards criteria derived for Ammonia, Nitrogen, Nitrate + -Nitrite, and TKN given that we observed a significant yet contradictory response in biological integrity for TKN and Nitrogen, Nitrate + Nitrite.

Phosphorus criteria seems to be related to biologically available ionic fractions rather than TP. Our NBI$_{TP}$ model was positively correlated with increasing biological integrity suggesting that we could not input enough Phosphorus into Indiana streams. Conversely, Chlorophyll a (phytoplankton), which is an early indicator measure, showed a strong negative relationship with increasing biological integrity. This relationship appears to be related to either the assimilative or mitigating capacity of the stream system. Total Phosphorus loading is apparently not significant as long as it can be biologically remediated in the stream system. However, once the biological available fraction exceeds the mitigating capacity of the system, Phytoplankton attains nuisance levels. Our SRI break point for Chlorophyll a (phytoplankton) is equivalent to the levels set to control summer Phytoplankton levels in Oregon (Dodds and Welch 2000).

5.8 Multidirectional Interpretation

Thus far we have discussed the application of the NBI using a unidirectional approach. Our models have been calibrated to detect significance in fish assemblage structure in relation to varying nutrient doses, and we have identified specific contaminant values which are predictive of these shifts. This approach places the model endpoint on contaminant prediction. However, the model could also be applied in reverse, where on observed contaminant concentration could be used to predict a biological expectation. Model sensitivity would likely not support prediction of specific species presence but could provide insight into expected fish assemblage patters and assemblage structure.

5.9 Management Implications: Ramification to Gulf of Mexico Hypoxia Loadings

Corn Belt loadings have revealed that high percentages of nutrients are emanating from less than 22% of the land mass (Smith et al. 2003) and principally come from Indiana and several other states. In order to achieve relief from nutrient impairment in streams and rivers, different land use management practices are needed to reduce the amounts of Nitrogen and Phosphorus entering the northern Gulf of Mexico.

The nutrient load to the northern Gulf of Mexico has been declining over the last decade as a result of public awareness and active management. The ramification of decreased loadings is that biological response has become more sensitive to shifts.

Our calibrated NBI models should show visible improvements in the management of hypoxia in the northern Gulf of Mexico since our models are based on specific response targets in biological assemblages. By targeting ecological shifts in biological integrity, these response shifts will improve biological integrity in concert with decreasing Nitrogen and Phosphorus levels. Differing shift response intervals can be phased in over time so that management objectives can be realized efficiently and without burdening the economy. The establishment of numerical nutrient criteria will show improvement in managing water resources in the United States; however, further actions will need to be implemented as threshold responses are attained. For example, incentives to producers in the agricultural industry must be advocated to encourage participation from the farming and livestock land managers. These incentives could be advocated through the farm bill and using capitalistic approaches such as a series of nutrient credits that can be applied to select land mass sizes. The implementation of nutrient criteria will require a staged management so that the influence of reduced fertilizers and nutrient credits can be determined without affecting water quality and quantity.

6 Conclusion

Nutrient loadings emanating from the Corn Belt have enriched surface waters and affected the biological integrity of aquatic assemblages including coastal waters in the Gulf of Mexico. Due to the complex interactions between the various forms of Nitrogen and Phosphorus within respective cycles, Total Nitrogen and Total Phosphorus cycling interactions can no longer be accepted as sole limiting factors in either marine or freshwaters. This study is conducted as part of the U.S. Environmental Protection Agency desire to development regional nutrient thresholds. The first objective of this study is to develop a biotic model capable of determining the contributions of various nutrients, including Nitrogen components and TP, in streams using fish assemblages. The second objective is to establish an approach for designating defensible nutrient biotic index score thresholds and corresponding nutrient concentrations, above which fish assemblages show alterations due to increasing nutrient concentrations. Sampling within Indiana's portion of the Corn Belt and Northern Great Plain Nutrient Ecoregion occurred from 1996 to 2007 at 1,274 sites. Nutrient data were reviewed for outliers and then sorted into three groups relative to drainage class. Each group was arranged into 15 ranges or "bins" using the Jenks optimization method in Arc GIS 9.3. Next, sites were assigned to each bin relative to observed concentrations. These bin assignments were used to populate the species occurrence model for nutrient optima calculation. Nutrient optima were calculated by dividing the sum of the weighted proportion of times a species occurred in each bin by the un-weighted proportion of times a species occurred in each bin. The derived nutrient optima were divided into eleven equal ranges, by nutrient, and tolerance scores (0–10) assigned with respect to each species derived optima. Nutrient tolerance scores were used to calculate Nutrient Biotic Index scores for each sampling site by summing the number of individuals of a given species at the site and multiplying times that species tolerance value then dividing by the total number of individuals at the site. A single break point was observed for unionized ammonia, which showed an $NBI_{Unionized\ Ammonia}$ score shift between 0.003 and 0.03 mg/L. The mean $NBI_{Unionized\ Ammonia}$ scores were 3.09 and 3.29, respectively. Nutrient Biotic $Index_{Unionized\ Ammonia}$ scores were significantly correlated with IBI score and IBI integrity class. Three break points were observed for Nitrogen, Nitrate + Nitrite, demonstrating a significant $NBI_{Nitrate+Nitrite}$ score shift at mean concentrations of 0.13, 1.09, 3.15 and 6.87 mg/L respectively. The mean $NBI_{Nitrate+Nitrite}$ scores were 5.58, 5.37, 5.82 and 6.25, respectively. The observed relationship produced a convex curve suggesting an enrichment signature. Nutrient Biotic $Index_{Nitrate+Nitrite}$ scores were significantly correlated with IBI score and IBI integrity class. Two break points were observed for Total Kjeldahl Nitrogen, which were significant. The mean concentrations of TKN were 0.4, 0.68, and 1.27, respectively. The mean NBI_{TKN} scores were 2.73, 3.10, and 3.37, respectively. Nutrient Biotic $Index_{TKN}$ scores were significantly related to IBI score and IBI integrity class. Two break points observed for TN were significant at concentrations of 0.56 mg/L and 3.30 mg/L. The mean NBI_{TN} scores were 4.60 and 4.85, respectively. Nutrient Biotic $Index_{TN}$ scores were not significantly related to IBI score or IBI integrity class. Two significant break points were observed for TP. The mean concentrations of TP were 0.07 and 0.32 mg/L, respectively and mean NBI_{TP} scores were 3.43 and 3.58, respectively. Nutrient Biotic $Index_{TP}$ scores were significantly related to IBI score and IBI integrity class. Two break points were observed for Chlorophyll a (periphyton), which were significant. Mean concentrations were 10.15 and 134.14 mg/m^2, respectively. Mean $NBI_{Periphyton}$ scores were 3.75 and 4.20, respectively. Nutrient Biotic $Index_{Periphyton}$ scores were not significantly related to IBI score, but were significantly related to IBI integrity class. Four break points were observed for Chlorophyll a (phytoplankton), which occurred at Chlorophyll a (phytoplankton) concentrations of 2.33, 10.98 and 49.13 µg/L, respectively. The mean $NBI_{Phytoplankton}$ scores were 3.43, 3.85 and 5.02, respectively. Nutrient Biotic $Index_{Phytoplankton}$ scores were significantly related to IBI score and IBI integrity class. Nutrient criteria concentration was interpreted for NBI and IBI integrity class relationships to establish protective nutrient concentration benchmarks. Proposed mean protection values are 3.0 µg/L for Unionized Ammonia, 130 µg/L for Nitrogen, Nitrate + Nitrite, 40 µg/L for TKN, 70 µg/L for TP, and 2.33 µg/L for Chlorophyll a (phytoplankton). Criteria established at or below these benchmarks should protect for both biological integrity of fish assemblages in Indiana as well as nutrient loadings into the Gulf of Mexico.

References

Alexander RB, Smith RA, Schwarz GE, Boyer EW, Nolan JV, Brakebill JW (2008) Differences in phosphorus and nitrogen delivery to the Gulf of Mexico from the Mississippi River basin. Environ Sci Technol 42:822–830

Arar EJ, Collins GB (1997) In vitro determination of chlorophyll a and phaeophytin *a* in marine and freshwater algae by fluorescence (rev. 1.2). U.S. Environmental Protection Agency, National Exposure Research Laboratory, Office of Research and Development, Cincinnati, pp. 1–22 (Method 445.0-1)

Augspurger T, Keller AE, Black MC, Cope WG (2003) Water quality guidance for protection of freshwater mussels (Unionidae) from ammonia exposure. Environ Toxicol Chem 22:2569–2575

Becker G (1983) Fishes of Wisconsin. University of Wisconsin Press, Wisconsin

Black RW, Munn MD, Plotnikoff RW (2004) Using macroinvertebrates to identify biota-land cover optima at multiple scales in the Pacific Northwest, USA. J N Am Benthol Soc 23:340–362

Boesch DF (2002) Challenges and opportunities for science in reducing nutrient over-enrichment in coastal ecosystems. Estuaries 25:886–900

Boyer EW, Goodale CL, Jaworski NA, Howarth RW (2002) Anthropogenic nitrogen sources and relationships to riverine nitrogen export in the northeastern USA. Biogeochemistry 57:137–169

Brakebill JW, Preston SD (2004) Digital data used to relate nutrient inputs to water quality in the Chesapeake Bay watershed, version 3.0. US Geological Survey, Washington (Water Resources Investigations Report 2004-1433)

Brandt D (2001) Temperature response regressions for 162 common macroinvertebrate taxa and supporting documentation. Idaho Department of Environmental quality, Idaho, pp 1–90

Buck S et al (2000) Nutrient criteria technical guidance manual, rivers and streams. EPA-822-B-00-002. United States Environmental Protection Agency, USA

Chutter FM (1972) An empirical biotic index of the quality of water in South African streams and rivers. Water Res 6:19–30

Dodds WK, Oaks RM (2004) A technique for establishing reference nutrient concentrations across watersheds affected by humans. Limnol Oceanogr 2:333–341

Dodds WK, Welch EB (2000) Establishing nutrient criteria in streams. J N Am Benthol Soc 19:186–196

Etnier DA, Starnes WC (1993) The fishes of Tennessee. The University of Tennessee Press, Knoxville

Environmental Protection Agency (EPA) (1998a) Level III ecoregions of the continental United States (Revision of Omernik 1987). US Environmental Protection Agency, Office of Water, Office of Science and Technology, USA

Environmental Protection Agency (EPA) (1998b) National strategy for the development of regional nutrient criteria. US Environmental Protection Agency, Office of Water, Office of Science and Technology, Washington (EPA 822-R-98-002)

Environmental Protection Agency (EPA) (1999) 1999 Update of Ambient Water Quality Criteria for Ammonia. US Environmental Protection Agency, Office of Water, Office of Science and Technology, Washington (EPA 822-R-99-014)

Gibson G, Carlson R, Simpson J, Smeltzer E, Gerritson J, Chapra S, Heiskary S, Jones J, Kennedy R (2000a). Nutrient criteria technical guidance manual—lakes and reservoirs. US Environmental Protection Agency, Office of Water, Washington (EPA-822-B00-001)

Gibson G, Carlson R, Simpson J, Smeltzer E, Gerritson J, Chapra S, Heiskary S, Jones J, Kennedy R (2000b) Nutrient criteria technical guidance manual—rivers and streams. US Environmental Protection Agency, Office of Water, Washington (EPA-822-B00-002)

Hilsenhoff WL (1987) An improved biotic index of organic stream pollution. Gt Lakes Entomol 20:31–39

Howarth RW, Marino R (2006) Nitrogen as the limiting nutrient for eutrophication in coastal marine ecosystems: evolving views over three decades. Limnol Oceanogr 31:364–376

Howarth RW, Sharpley A, Walker D (2002) Sources of nutrient pollution to coastal waters in the United States: implications for achieving coastal water quality goals. Estuaries 25:656–676

Indiana Department of Environmental Management (1992) Standard operating procedures for electrofishing. IDEM, Indianapolis

Indiana Department of Environment Management (2004) Quality assurance project plan for Indiana Surface Water Quality Monitoring Programs, revision 3. 100/29/338/073/2004. Indiana Department of Environmental Management, Office of Water Quality, Assessment Branch, Indianapolis

Jenks GF (1977) Optimal data classification for choropleth maps. Occasional paper No. 2. University of Kansas, Department of Geography, Lawrence

Jongman RHG, ter Braak CFJ, van Tongeren OFR (1987) Data analysis in community and landscape ecology. Pudoc Wageningen, Netherlands

Kaushal SS, Lewis WM Jr (2005) Fate and transport of dissolved organic nitrogen in minimally disturbed montane streams of Colorado, USA. Biogeochemistry 74:303–321

Leopold LB, Woolman MG, Miller JP (1964) Fluvial processes in geomorphology. W.H. Freeman, San Francisco

Lowe BS, Leer DR, Frey JW, Caskey BJ (2008) Occurrence and distribution of algal biomass and its relation to nutrients and basin characteristics in Indianan streams. U.S. Geological Survey Scientific Investigation Report 2008-5203, pp 1–146

Mississippi River/Gulf of Mexico Nutrient Task Force (2004) A science strategy to support management decisions related to hypoxia in the Northern Gulf of Mexico and excess nutrients in the Mississippi River Basin. Monitoring, Modeling, and Research Workgroup of the Mississippi River/Gulf of Mexico Watershed Nutrient Task Force, U.S. Geological Survey Circular 1270

Morris CC, Simon TP, Newhouse SA (2006) A local in situ approach for stressor identification of biologically impaired aquatic systems. Arch Environ Contam Toxic 50:325–334

Moulton SR II, Kennen JG, Goldstein RM, Hambrook JA (2002) Revised protocols for sampling algal, invertebrate, and fish communities as part of the National Water-Quality Assessment Program. U.S. Geological Survey Open-File Report 02-150, pp 1–75

Omernik JM (1976) The influence of land use on stream nutrient levels. US Environmental Protection Agency, Corvallis (Ecological Research Series, EPA-600/3-76-014)

Omernik JM (1977) Nonpoint source-stream nutrient level relationships: a nationwide study. US Environmental Protection Agency, Corvallis (Ecological Research Series, EPA-600/3-77-105)

Rabalias NN, Turner RE, Dortch Q, Justic D, Bierman VJ, Weisman WJ Jr (2002) Nutrient enhancement productivity in the northern Gulf of Mexico: past, present, and future. Hydrobiologia 475/476:39–63

Reckhow KH, Arhonditsis GB, Kenney MA, Hauser L, Tribo J, Wu C, Elcock KJ, Steinberg LJ, Stow CA, McBride SJ (2005) A predictive approach to nutrient criteria. Environ Sci Technol 39:2913–2919

Scavia D, Donnelly KA (2007) Reassessing hypoxia forecasts for the Gulf of Mexico. Environ Sci Technol 41:8111–8117

Shelton LR (1994) Field guide for collecting and processing stream-water samples for the National Water-Quality Assessment Program. U.S. Geological Survey Open-File Report 94-455, pp 1–646

Simon TP, Morris CC (2009) Biological response signature of oil brine threats, sediment contaminants, and crayfish assemblages in an Indiana watershed, USA. Arch Environ Contam Toxic 56:96–110

Smith RA, Alexander RB, Schwarz GE (2003) Natural background concentrations of nutrients in streams and rivers of the conterminous United States. Environ Sci Technol 37:3039–3047

Smith AJ, Bode RW, Kleppel GS (2007) A nutrient biotic index (NBI) for use with benthic macroinvertebrate communities. Ecol Indic 7:371–386

StatSoft (2007) STATISTICA for Windows. StatSoft, Tulsa

Stoddard JL, Larsen DP, Hawkins CP, Johnson RK, Norris RH (2006) Setting expectations for the ecological condition of streams: the concept of reference condition. Ecol Appl 16:1267–1276

Ter Braak JF, Juggins F (1993) Weighted averaging partial least squares regression (WA-PLS): an improved method for reconstructing environmental variables from species assemblages. Hydrobiologia 269/270:485–502

Turner RE, Rabalais RR, Justic D (2008) Gulf of Mexico hypoxia: alternate states and a legacy. Environ Sci Policy 42:2323–2327

Wollheim WM, Pellerin BA, Vorosmarty CJ, Hopkinsion CS (2005) Nitrogen retention in urbanizing headwater catchments. Ecosystems 8:871–884

Appendix

C. C. Morris and T. P. Simon, *Nutrient Indicator Models for Determining Biologically Relevant Levels*, SpringerBriefs in Environmental Science, DOI: 10.1007/978-94-007-4129-4, © The Author(s) 2012

Scientific name	Common name	Drainage class % occurrence			Tolerance score						
		Small	Medium	Large	Unionized ammonia	Nitrate-nitrite	TKN	TN	TP	Phyto.	Peri.
Lepisosteidae											
Lepisosteus osseus	Longnose gar	–	10	29	7	5	8	7	5	2	2
L. platostomus	Shortnose gar	–	–	40	6	1	5	4	10	6	4
Clupeidae											
Dorosoma cepedianum	Gizzard shad	–	41	98	4	3	6	6	5	5	4
Esocidae											
Esox americanus	Grass pickerel	15	25	–	3	7	5	6	4	4	5
Umbridae											
Umbra limi	Central mudminnow	13	–	–	0	4	1	2	1	1	4
Cyprinidae											
Cyprinus carpio	Common carp	–	63	86	4	3	5	6	6	6	3
Hybognathus nuchalis	Mississippi silvery minnow	–	–	32	1	6	5	7	3	6	3
Semotilus atromaculatus	Creek chub	83	38	–	3	7	2	5	2	4	4
Rhinichthys obtusus	Western blacknose dace	31	–	–	0	8	0	6	4	0	4
Nocomis micropogon	River chub	–	16	–	0	10	3	10	2	0	9
N. biguttatus	Hornyhead chub	9	10	–	0	5	3	4	1	0	10
Notropis rubellus	Rosyface shiner	–	20	–	1	7	0	8	2	1	5
N. atherinoides	Emerald shiner	–	–	87	4	3	5	6	4	7	3
N. stramineus	Sand Shiner	16	41	28	1	8	4	7	5	–	4
N. volucellus	Mimic shiner	–	10	–	–	–	4	5	2	0	4
N. photogenis	Silver shiner	–	17	–	1	7	4	8	3	1	3
Ericymba buccata	Silverjaw minnow	29	21	–	0	7	1	5	1	0	5
Hybopsis amblops	Bigeye chub	–	20	–	0	6	0	7	2	1	4
Phenacobius mirabilis	Suckermouth minnow	6	16	–	0	4	2	4	1	5	4

(continued)

(continued)

Scientific name	Common name	Drainage class % occurrence			Tolerance score						
		Small	Medium	Large	Unionized ammonia	Nitrate-nitrite	TKN	TN	TP	Phyto.	Peri.
Campostoma anomalum	Central stoneroller	56	36	–	3	7	0	4	3	5	6
C. oligolepis	Largescale stoneroller	8	–	–	3	4	1	0	3	0	0
Pimephales notatus	Bluntnose minnow	68	76	51	4	7	5	5	3	4	4
P. promelas	Fathead minnow	10	–	–	10	9	0	10	1	1	8
P. vigilax	Bullhead minnow	–	–	79	4	3	5	7	6	5	6
Phoxinus erythrogaster	Southern redbelly dace	7	–	–	0	9	0	7	10	0	0
Cyprinella spiloptera	Spotfin shiner	21	76	86	3	4	3	5	4	5	4
C. whipplei	Steelcolor shiner	–	26	34	6	3	4	6	4	1	3
Luxilus chrysocephalus	Striped shiner	39	46	–	5	6	2	4	3	1	4
Lythrurus umbratilis	Redfin shiner	16	–	–	1	8	0	6	2	2	6
Catostomidae	–	–	–	–	–	–	–	–	–	–	–
Catostomus commersoni	White sucker	56	40	–	3	6	4	5	3	3	5
Carpiodes cyprinus	Quillback	–	23	28	5	8	6	6	9	5	2
C. carpio	River carpsucker	–	–	90	3	3	5	5	6	7	3
Erimyzon oblongus	Creek chubsucker	13	–	–	0	3	1	0	6	2	6
Moxostoma macrolepidotum	Shorthead redhorse	–	33	53	4	9	4	7	8	5	3
M. anisurum	Silver redhorse	–	24	27	0	3	1	3	9	8	1
M. duquesnei	Black redhorse	–	44	–	1	7	3	7	7	1	5
M. erythrurum	Golden redhorse	–	67	37	6	5	5	7	7	4	4
Hypentelium nigricans	Northern hogsucker	25	60	–	1	5	1	3	3	3	3
Cycleptus elongatus	Blue sucker	–	–	42	4	4	3	6	4	8	5
Ictiobus bubalus	smallmouth buffalo	–	14	72	5	5	5	5	4	5	2

(continued)

(continued)

| Scientific name | Common name | Drainage class % occurrence | | | Tolerance score | | | | | | |
		Small	Medium	Large	Unionized ammonia	Nitrate-nitrite	TKN	TN	TP	Phyto.	Peri.
I. cyprinellus	Bigmouth buffalo	–	–	39	5	0	4	2	9	8	3
Ictiobus niger	Black buffalo	–	–	29	0	10	1	4	8	9	2
Minytrema melanops	Spotted sucker	–	30	24	5	2	6	4	2	5	6
Ictaluridae	–	–	–	–	–	–	–	–	–	–	–
Ictalurus punctatus	Channel catfish	–	40	97	4	4	5	5	7	6	4
Noturus flavus	Stonecat	–	21	–	2	5	1	5	0	1	5
Noturus miurus	Brindled madtom	–	12	–	1	0	0	0	0	1	3
Pylodictus olivaris	Flathead catfish	–	22	86	4	2	5	4	7	4	4
Ameiurus melas	Black bullhead	6	–	–	1	10	10	10	8	–	–
A. natalis	Yellow bullhead	37	34	–	6	5	4	5	6	2	3
Fundulidae	–	–	–	–	–	–	–	–	–	–	–
Fundulus notatus	Blackstripe topminnow	23	–	–	2	6	5	4	3	2	7
Poeciliidae	–	–	–	–	–	–	–	–	–	–	–
Gambusia affinis	Western mosquitofish	6	–	–	0	1	2	1	5	–	–
Atherinidae	–	–	–	–	–	–	–	–	–	–	–
Labidesthes sicculus	Brook silverside	–	16	–	5	0	10	3	1	2	2
Cottidae	–	–	–	–	–	–	–	–	–	–	–
Cottus bairdi	Mottled sculpin	19	22	–	1	8	1	8	4	3	4
Centrarchidae	–	–	–	–	–	–	–	–	–	–	–
Ambloplites rupestris	Rock bass	24	70	–	2	6	4	4	6	1	4
Lepomis cyanellus	Green sunfish	65	74	40	3	4	5	5	3	6	3
L. gulosus	Warmouth	5	12	–	10	0	10	1	1	2	5
L. macrochirus	Bluegill	45	80	79	4	4	5	4	4	5	3
L. megalotis	Longear sunfish	41	78	89	4	4	4	4	4	4	3
Micropterus dolomieu	Smallmouth bass	–	54	55	4	5	3	6	8	4	4

(continued)

(continued)

Scientific name	Common name	Drainage class % occurrence			Tolerance score						
		Small	Medium	Large	Unionized ammonia	Nitrate-nitrite	TKN	TN	TP	Phyto.	Peri.
M. salmoides	Largemouth bass	18	35	24	3	4	4	3	3	9	2
M. punctulatus	Spotted bass	22	55	87	3	3	3	3	4	4	3
Pomoxis nigromaculatus	Black crappie	–	20	25	10	2	5	3	3	6	2
Percidae	–	–	–	–	–	–	–	–	–	–	–
Etheostoma spectabile	Orangethroat darter	39	21	–	4	7	3	7	3	5	4
E. nigrum	Johnny darter	59	52	–	4	5	3	4	3	5	4
E. blennioides	Greenside darter	23	49	–	1	6	2	5	3	3	3
E. caeruleum	Rainbow darter	25	45	–	2	5	0	4	4	0	5
E. flabellare	Fantail darter	24	21	–	3	4	2	2	1	1	3
Percina caprodes	Logperch	–	45	–	3	4	3	4	2	3	4
P. sciera	Dusky darter	–	29	52	5	5	6	6	7	6	4
P. maculata	Blackside darter	8	25	–	3	2	7	1	3	10	4
P. phoxocephala	Slenderhead darter	–	16	28	4	3	4	5	1	2	7
Sander canadense	Sauger	–	–	36	6	8	3	9	10	4	5
Sciaenidae	–	–	–	–	–	–	–	–	–	–	–
Aplodinotus grunniens	Freshwater drum	–	23	84	4	3	5	4	4	5	3

Index

C. C. Morris and T. P. Simon, *Nutrient Indicator Models for Determining*   67
Biologically Relevant Levels, SpringerBriefs in Environmental Science,
DOI: 10.1007/978-94-007-4129-4, © The Author(s) 2012